A New Era of Discovery: Unveiling the Potential of Quantum Materials for Revolutionizing Technology

Mourine

Contents

1. Introduction

In the era of discovery, explorers changed the old world with their travels towards unknown parts of the globe. Around five hundred years later, the era of information was put in motion by the digital revolution, founded in the development of the semiconductor technology. Nowadays, the semiconductor industry is mature and new kinds of technologies are expected to exceed them. Quantum materials are found next in line to continue and elevate semiconductors [1]. In these materials, electrons cannot be considered as independent particles any more, due to strong interactions. Novel phases emerge as a consequence of these collective phenomena. This results in various quantum phases manifesting macroscopically, of which superconductivity is the best example.

The challenges of condensed matter physics are to find novel quantum phases with potential industrial implantation. The success of this quest would be founded in the understanding of the electron-electron Coulomb interaction and its relationship with the spin and spatial component. In this quest, phase diagrams need to be characterised, they occupy the role of maps on which the frontiers of the world were reshaped by adventurous explorers.

A key challenge in the understanding of these quantum materials phase diagrams is to study the spatial distribution of their emergent phases. The interplay of short-range and long-range electronic interactions can result in phase-separated physical regions such as spin-density wave and superconducting [2–4]. This is especially interesting when more than one phase is close to instability and competes producing spatial variation. A direct method to address this challenge is scanning tunnelling microscopy (STM). An imaging technique which locally proves the electronic density of states.

In this book, the correlated electronic phases of two quantum material are investigated by spectroscopy-imaging scanning tunnelling microscopy: Sr_2IrO_4, a spin-orbit assisted Mott insulator, and Li doped NaFeAs, a Fe-based superconductor. In both materials, magnetic order is found at low temperatures before introducing any doping. However, the magnetic order is reached by two different ways: Sr_2IrO_4 is a Mott insulator at half-filling, the electronic correlations create antiferromagnetic order. In the case of $Na_{1-x}Li_xFeAs$, the magnetic order is driven by a spin density wave instability. the changes in the phase diagrams of these compounds are investigated in this book.

This work is structured as follows: In the second chapter (2), the physical concepts touched in this **book** are summarized. Instabilities of the Fermi sea due to electron-phonon and electron-electron interactions are reviewed: charge and spin density waves and superconductivity. Due to the importance of the electron-electron interactions, the Hubbard model is introduced, with a focus in the half-filling limit. The scope of this chapter is to provide a foundation for understanding the main results presented in the subsequent sections.

An introduction to STM is given in the third chapter (3). It covers the quantum theory of tunnelling and its application to STM. The main operation methods are explained for topographic and spectroscopic data, with an emphasis in spectroscopy-imaging STM and Fourier transform spectroscopy. The quasiparticle interference analysis is described. Furthermore, this chapter contains a description of the experimental setup used in the present work.

The fourth chapter (4) is focused on the electronic properties of Sr_2IrO_4, a spin-orbit assisted Mott insulator. The chapter starts with an introduction to the physics of Iridium oxides compounds (Iridiates). It continues with a description of the studied sample. Crystal growth, bulk characterization and tunnelling measurements are described. This chapter has two main results: The spin-polaron quasiparticle ladder spectrum is revealed in the positive side of the STS spectrum and the Mott-insulator transition is studied with a percolative charge transfer model.

Li doped NaFeAS, newly member of the iron-based superconductors, is studied in the fifth chapter (5). The topographic and spectroscopic features of the $Na_{1-x}Li_xFeAs$ samples are characterized. The work done here discovered a charge instability in the system and studied its connection to nematic order. The charge instability is identified in real and momentum space by spectroscopy imaging and Fourier transform STS. The results are compared with the experimental ARPES and NRM data.

Finally, the main results and findings of this **book** are summarized and presented in the sixth chapter (6).

2. Foundations

This chapter introduces the main physics concepts underlying this **book**.

The concepts discussed here are:

- Charge and spin density waves.

- Electronic correlation based on the Hubbard model and the $t - J$ model: Mott insulator state and spin polaron.

- Conventional and unconventional superconductivity, with the consequences for the order parameter.

A deeper analysis of the physics of the material studied in this **book** will be given in the correspondent chapters: Chapter 4 for Sr_2IrO_4 and Chapter 5 for Li doped NaFeAs.

2.1. Density waves

Density waves are broken symmetry states triggered by electron-phonon or electron-electron interactions. The ground state is that charge (charge density wave, CDW) or spin (spin density wave, SDW) are ordered displaying a periodic spatial modulation. The periodic modulation of the electron/spin density is determined by the Fermi wave vector k_F [5].

2.1.1. Density wave instability

In the conventional picture density waves arise from a instability of the Fermi surface. The presence of such instability can be understood when looking at the response of the electron gas to an external perturbation. This is described by the Lindhard function. The equation for the d-dimensional Lindhard functions is:

$$\chi^0(\mathbf{q}) = \int \frac{d\mathbf{k}}{(2\pi)^d} \frac{f_k - f_{k+q}}{\epsilon_k - \epsilon_{k+q}} \tag{2.1.1}$$

Here, $f_k = f_{\epsilon_k}$ is the Fermi function and ϵ_k describes the band structure the system. As shown in Fig. 2.1.1 in 1D $\chi^0(\mathbf{q})$ diverges at $q = 2k_F$. This picture can be extended to

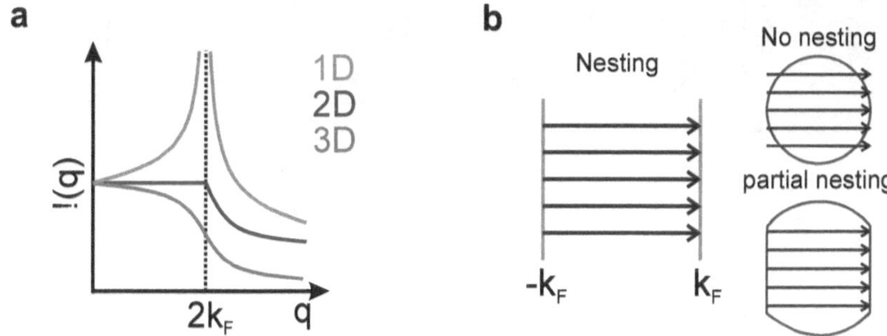

Figure 2.1.1.: (a) the response of the Lindhard function in 1 (green), 2 (blue) and 3 dimensions (red). For a one-dimensional (1D) $\chi(\mathbf{q})$ diverges at $q = 2\,k_F$. In three dimensions, $\chi(\mathbf{q})$ decreases with increasing $\mathbf{q}$ and the derivative has a logarithmic singularity at $q = 2\,k_F$. (b) Fermi surface nesting, partial nesting and no nesting in 2D.

higher dimensions with the concept of Fermi surface nesting (Fig. 2.1.1 (b)). The nesting conditions are fulfilled for particular topologies of the Fermi surface. It is called nesting of the Fermi surface when parallel regions of the Fermi surface are connected to each other by a single vector $\mathbf{q}$, known as nesting vector, in this case, the energies ϵ_k and ϵ_{k+q} are degenerate. In some cases, Fermi surface nesting can lower the energy of the system.

2.1.2. Charge density wave instability

In the simplest case of CDW in metals the ground state develops as a consequence of electron-phonons interactions in low dimensional metals. The mechanism behind a CDW was first proposed by Fröhlich [6] and Peierls [7]. The CDW instability is a consequence of the Fermi surface nesting that produces a lattice distortion. However other mechanisms are also possible. A second type of CDW is produced with no Fermi surface nesting. In this case, the lattice and electronic instabilities are tied to each other [8]. A more interesting scenario for this **book** is produced when the electron-electron interactions are strong enough to drive the CDW instability. In these systems, charge order can compete or coexist with magnetism and unconventional superconductivity [9–11].

At the instability, the phonon frequency (w_{2k_F}) is renormalized to have lower energy (Kohn anomaly). This renormalization is strongly dependent on the temperature.

$$\omega^2_{ren\,2k_F} = w^2_{2k_F} + \frac{2g^2 n(\epsilon_F)\omega_{2k_F}}{\hbar} \ln\left(\frac{1.14\epsilon_0}{k_B T}\right) \tag{2.1.2}$$

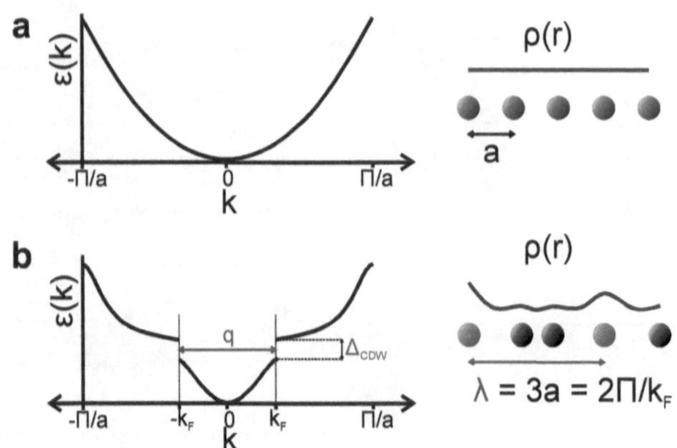

Figure 2.1.2.: The electronic density of states in a 1D metal in real ($\rho(r)$) and momentum ($\epsilon(k)$) space. (a) Single-band dispersion and atomic lattice with lattice constant a. (b) Opening of a gap in momentum space produced by a CDW modulation of the electron density (with wavelength $\lambda = \pi/k_F$) and lattice distortion (usually quite small, $\leq 1\%$ of a).

With k_B the Boltzmann constant, $\hbar$ the Plank constant and g the electron-phonon coupling. The critical temperature, T_{CDW}, drives the renormalized phonon frequency to zero and freezes the lattice distortion.

$$k_B T_{CDW} = 1.14\epsilon_0 e^{-1/\lambda} \tag{2.1.3}$$

with $\lambda = \frac{g^2 n(\epsilon_F)}{\hbar \omega_{2k_F}} = g\prime n(\epsilon_F)$. The phase transition is defined as the temperature where $\omega_{ren\,2K_F} \mapsto 0$ due to the strong divergence response of $\chi(\mathbf{q}, T)$. After the transition a gap opens at k_F, which leads to a lowering of the energy (see Fig. 2.1.2). In 1D and weak coupling limit $E_{normal} - E_{CDW} = \frac{n(\epsilon_F)}{2}\Delta^2$:

$$2\Delta = 3.52 k_B T_{CDW} \tag{2.1.4}$$

2.2. Electronic correlations and Hubbard model

"Electronic correlations" refers to the interactions between two or more electrons in a solid via Coulomb repulsion. The electron-electron correlation, which produce attractive and repulsive forces, are responsible for new ground states or complex collective behaviour in condensed matter physics. The Hubbard model plays a central role in the understanding of correlated electron systems. The beauty of the Hubbard model is the simplicity of the

Hamiltonian which allows a meaningful description of two opposing tendencies:

- The kinetic energy (hopping) promotes delocalization of the electrons into itinerant states (Bloch waves), leading to metallic behaviour.

- The electron-electron interaction strengthens localization of the electrons state, driving the transition to a Mott insulator.

In the one-band Hubbard model the Hamiltonian is given by:

$$H = -t \sum_{\langle \mathbf{j},\mathbf{l} \rangle} \sum_{\sigma} \left(c_{\mathbf{j}\sigma}^{\dagger} c_{\mathbf{l}\sigma} + c_{\mathbf{l}\sigma}^{\dagger} c_{\mathbf{j}\sigma} \right) + U \sum_{\mathbf{j}} \hat{n}_{\mathbf{j}\uparrow} \hat{n}_{\mathbf{j}\downarrow} \tag{2.2.1}$$

Here $c_{\mathbf{j}\sigma}^{\dagger}$ creates an electron with spin σ at the lattice site with index $\mathbf{j}$. $\hat{n}_{\mathbf{j}\sigma} = c_{\mathbf{j}\sigma}^{\dagger} c_{\mathbf{j}\sigma}$ is the corresponding occupation number. Parametrized by t, the sum over nearest neighbour pairs $\langle \mathbf{j},\mathbf{l} \rangle$ describes the kinetic energy of the electrons H_{band} (normally the tight-binding band is chosen as the unperturbed part). The interaction term with the Hubbard interaction parameter U, describes the Coulomb repulsion between electrons sharing the same orbital H_U. Where the Hubbard interaction parameter U is:

$$U = \int d\mathbf{r}_1 \int d\mathbf{r}_2 \, |\phi(\mathbf{r}_1 - \mathbf{R}_{\mathbf{j}})|^2 \, \frac{e^2}{|\mathbf{r}_1 - \mathbf{r}_2|} \, |\phi(\mathbf{r}_2 - \mathbf{R}_{\mathbf{j}})|^2 \tag{2.2.2}$$

A realistic Hamiltonian should contain lots of inter-site terms. The success of the Hubbard model comes from the easiness of describing the competition between metallic behaviour and the Coulomb energy.

2.2.1. Mott insulator

An important case of the Hubbard model is the Mott insulator found at half filling. In the case that the orbitals of each atomic site can be twofold occupied, with one electron having spin-up and one having spin-down. When the material is in the half-filled limit and there are as many electrons as lattice sites, the energy is minimized when the electrons stay localized one on each site. This is the ground state for $U = \infty$. When U is large but finite, each site remains singly occupied. However, an electron can shortly jump onto the next neighbour site. This jump must respect Pauli exclusion principle, therefore their spins have to be antiparallel. The system can save energy by aligning the spins antiparallel and therefore creating antiferromagnetic order. Pauli exclusion principle also prevents the electrons from hopping to the next-nearest neighbour. The magnetically ordered phase is also an insulator, with large U and one electron per site is known as Mott insulator.

Associated to a Mott insulator are the concepts of upper and lower Hubbard bands. It helps to imaging the excitation spectrum of one electron shown in Fig. 2.2.1 (a) with ground

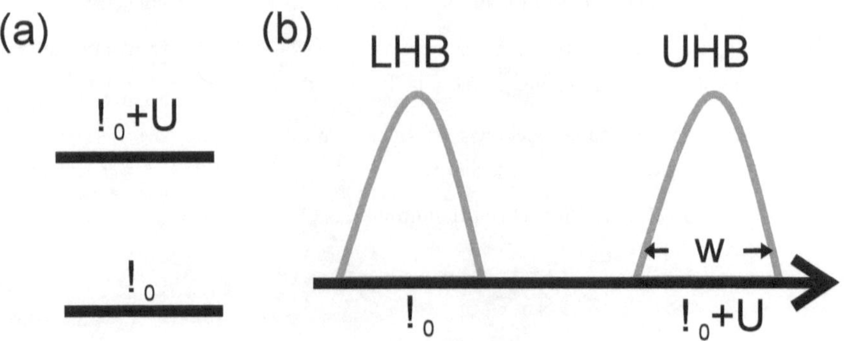

Figure 2.2.1.: (a) Two levels spectrum with ground state $\epsilon_0 = \epsilon_{at}$ (B) Sketch of the upper and lower Hubbard bands (LHB and UHB, respectively).

state $\epsilon_0 = \epsilon_{at}$ and excited state $\epsilon_{ex} = \epsilon_{at} + U$. Inside the AFM ground state, the electron can hop creating a band with bandwidth W ($W = 2zt$ where z is the coordination number for the hypercubic lattice) centred at $\epsilon_{at} + U$ as sketched in Fig. 2.2.1 (b). The splitting of these two Hubbard subbands is a correlation effect. The concept of the Hubbard subbands helps to understand the insulator to metal transition in terms of hopping and correlation energies. In the case of half filling, the metal-insulator transition occurs at the crossover when $U \sim W$. And the system becomes metallic when U is smaller than the electronic bandwidth.

2.2.2. t-J model

When extra charges are introduced into the system, it is displaced from the half-filling limit, which has important repercussions. An important consequence is the emergence of superconductivity [12]. However the many-body Hamiltonian is extremely complicated. The Hubbard model at the strong coupling limit ($U/t >> 1$) and at half filling can be reduced to the famous $t - J$ effective model. This model gives valuable insights on the nature of the correlated motion of electrons. At this limit the broadening of the atomic levels results in the Hubbard subbands. At large U the system is well into the insulating phase. In this case the movement of charges in the antiferromagnetic background is governed by the correlations of the antiparallel nearest-neighbour spin, therefore by Pauli exclusion principle. In the strong coupling limit, it is possible to treat the hopping part of the Hamiltonian, h_{band} as a perturbation. The so-called t-J model is derived using the

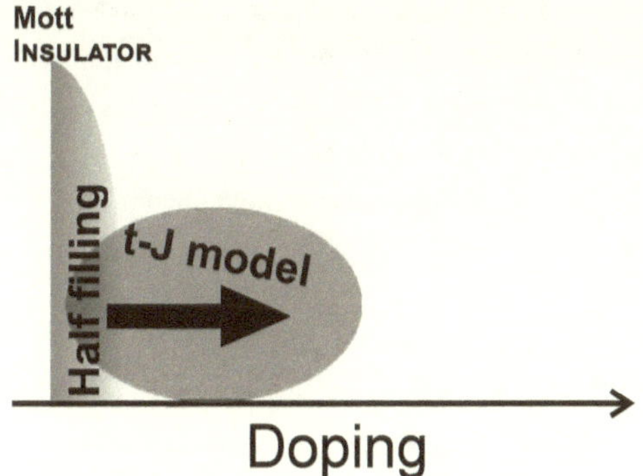

Figure 2.2.2.: Hubbard model phase diagram. At half filling the system develops a Mott insulator ground states.

second order of the perturbation theory:

$$H_{t-J} = -t \sum_{\langle \mathbf{i,j} \rangle, \sigma} \left[(1 - \hat{n}_{\mathbf{i}-\sigma}) c_{\mathbf{i}\sigma}^{\dagger} c_{\mathbf{j}\sigma} (1 - \hat{n}_{\mathbf{j}-\sigma}) + H.c. \right] + J \sum_{\langle \mathbf{i,j} \rangle} \left[\mathbf{S_i} \cdot \mathbf{S_j} - \frac{\hat{n}_{\mathbf{i}} \hat{n}_{\mathbf{i}}}{4} \right] \qquad (2.2.3)$$

Where the first term of the Hamiltonian describes the hopping constrained to a subspace with no doubly occupancy. It is existentially the hopping of hole or empty sites. The nearest-neighbour spins interact antiferromagnetically, with coupling constant $J = \frac{4t^2}{U}$, $\mathbf{S}_i = c_{i\alpha}^{\dagger} \boldsymbol{\sigma}_{\alpha\beta} c_{j\beta}/2$ and $\boldsymbol{\sigma} = (\sigma^x, \sigma^y, \sigma^z)$ the Pauli matrices.

For half filling, the model is reduced to the well-known antiferromagnetic Heisenberg model, with kinetic exchange Hamiltonian:

$$H_{AFM} = J \sum_{\langle \mathbf{i,j} \rangle} \left(\mathbf{S_i} \cdot \mathbf{S_j} - \frac{\hat{n}_{\mathbf{i}} \hat{n}_{\mathbf{i}}}{4} \right) \qquad (2.2.4)$$

Here the charge degrees of freedom are integrated out. This happens because the interactions are so strong that they localize the electrons. Thus only the virtual hopping produces an interaction between the spins. The strongly correlated electrons have been replaced by spin degrees of freedom. Therefore, the dimensionality of the problem has been largely reduced.

One of the biggest success of the $t - J$ model is the description of the problem of a hole moving in an antiferromagnetic background. The model is of great importance in condensed matter physics due to its relation with a doped Mott insulator. This is because, in unconventional superconductors the parent compounds acquire superconductivity through doping, which is understood as charge moving in the AFM planes. In fact, some experimental observation of unusual properties of the cuprates are naturally explained within the $t - J$ model, such as phase separation, d-wave superconductivity or a charge density instability. Extensive reviews on the physics of cuprates starting from doping a Mott insulator are written by Dagotto [12] Chernyshev and Wood [13], Lee et al. [14] and Ogata et al. [15].

t-J_z model and Spin-polaron dispersion

In the Heisenberg model low excitations are created by flipping one spin. These excitations are called spin waves and propagate through the lattice. The quasiparticle associated to the spin waves are called magnons. Just as phonons are quantized lattice waves, magnons are quantized magnetization density waves. The $t - J$ model can be reduced to the simpler $t - J_z$ model when spin fluctuations are neglected. Consequently, the ground state can be described by a classical Néel sate. Here the motion of a hole in the antiferromagnetic background can be described by an anisotropic $t - J$ model in the Ising limit.

When a charge carrier is moving in a magnetic medium, it is accompanied by deviations of localized spins. The distortion of the magnetic background can be described as a strong coupling of the hole and the spin degrees of freedom, in similar manner to the electron-phonon polaron case. This magnetic quasiparticle is called spin-polaron.

When fluctuations are neglected ($t - J_z$ model) the spin-polaron is confined. This confinement is explained in the panels of Fig. 2.2.3. Here, a hole moving in the antiferromagnetic background is constrained to oscillate around its origin with a string-like motion (as introduced by Bulaevskii et al. [16]). When the hole jumps it creates a region of ferromagnetically aligned spins which will have an attractive interaction with the hole (shadow region in Fig. 2.2.3 (b-d)). This interaction will result in an elastic potential energy. The energy of the string will be proportional to the length of the hopping path (L), $E_{string} \propto JL$. Therefore, the hole will be effectively confined in the string potential. The excitation spectrum consists in dispersionless excitations separated by energy intervals known as ladder spectrum [17] as shown in

When the spin fluctuations are allowed, they can flip one spin of the chain. Thus, the antiferromagnetic background can be healed. Consequently, the spring potential is reduced and therefore, the spin-polaron is allowed to propagate. So, the quasiparticle becomes dispersive and compromised the ladder structure.

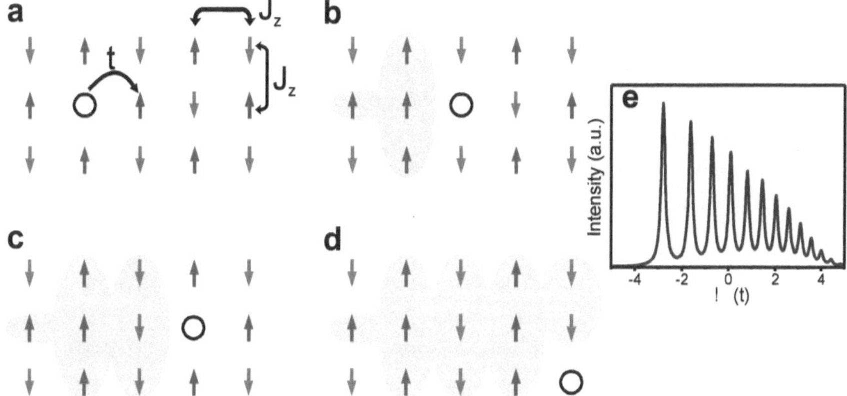

Figure 2.2.3.: The $t - J_z$ model which describes a hole in an antiferromagnetic background with the kinetic energy of the hole t and J_z the antiferromagnetic exchange coupling energy between neighbouring spins. The panels (a-d) show how the hopping of one hole conserves the spin and therefore creates a region where the antiferromagnetic correlations are destroyed and the magnetic energy is increased. The magnetic energy acts as an elastic potential and confines the hole to its original position.

2.3. BCS theory of Superconductivity

Superconductivity is a pure quantum phenomenon that can be recognized macroscopically. A material is considered as superconductor when it loses its electrical resistivity under a certain temperature, known as transition temperature (T_c) and the material exhibits perfect diamagnetism, this is to expel any external magnetic field. The second is known as Meissner effect and is the defining probe for superconductivity.

Superconductivity was first observed in H. Kamerlingh Onnes's lab, when the electrical resistance of mercury dropped to zero at 4.19 K. The microscopic theoretical understanding for the phenomena of superconductivity was achieved in the decade of the 1950's. Led by the effort of J. Bardeen, L. N. Cooper and J. R. Schrieffer [18]. Their microscopic theory of superconductivity is now known as BCS theory of superconductivity. It is a mean field theory based on the instability of the free electron due to the formation of correlations between states with opposite momenta and spins (so-called Cooper pairs). The Fermi sea is unstable against the formation of Cooper pair and can condense into a ground state where all electrons are paired. The Cooper instability is produced by electron-electron interactions mediated by phonon.

The attractive interaction can be understood thinking in an electron propagating through a crystal lattice. While the electron propagates it attracts the positive ions. Thus the electron creates a effective positive trace behind him. This effective trace is then felt by other electrons as an attractive interaction.

The BCS Hamiltonian can be written as:

$$\hat{H}_{BCS} = \sum_{\mathbf{k}\sigma} \epsilon_{\mathbf{k}} n^{\dagger}_{\mathbf{k}\sigma} n_{\mathbf{k}\sigma} + \sum_{\mathbf{k}\mathbf{k}\prime} V_{\mathbf{k}\mathbf{k}\prime} b^{\dagger}_{\mathbf{k}} b_{\mathbf{k}\prime} \tag{2.3.1}$$

Where $b^{\dagger}_{\mathbf{k}} = n^{\dagger}_{\mathbf{k}\uparrow} n^{\dagger}_{-\mathbf{k}\downarrow}$ and $b_{\mathbf{k}} = n_{-\mathbf{k}\downarrow} n_{\mathbf{k}\uparrow}$ are the creation and annihilation operators of the Cooper pairs.

After the mean field approximation:

$$H^{MF}_{BCS} = \sum_{\mathbf{k}\sigma} \epsilon_{\mathbf{k}} n^{\dagger}_{\mathbf{k}\sigma} n_{\mathbf{k}\sigma} - \sum_{\mathbf{k}} \Delta_{\mathbf{k}} b^{\dagger}_{\mathbf{k}} - \sum_{\mathbf{k}} \Delta^{*}_{\mathbf{k}} b_{\mathbf{k}\prime} \tag{2.3.2}$$

where $\Delta_{\mathbf{k}}$ is the BCS order parameter.

The Hamiltonian can be solved with the so-called Bogoliubov transformation and derive the Bogoliubov quasiparticles:

$$|u_{\mathbf{k}}|^2 = \frac{1}{2}\left(1 + \frac{\epsilon_{\mathbf{k}}}{E_{\mathbf{k}}}\right) \; ; \; |\nu_{\mathbf{k}}|^2 = \frac{1}{2}\left(1 - \frac{\epsilon_{\mathbf{k}}}{E_{\mathbf{k}}}\right) \tag{2.3.3}$$

with energy dispersion:

$$E_{\mathbf{k}}^2 = \epsilon_{\mathbf{k}}^2 + |\Delta_{\mathbf{k}}|^2 \tag{2.3.4}$$

The Bogoliubov quasiparticles will have a finite minimal energy $\Delta_0 = \Delta_{\mathbf{k}_F}$, known as the energy gap. As consequence of the energy gap, there are no Bogoliubov quasiparticles for energies below Δ_0. The gap function can be defined at zero temperature as $\Delta_{\mathbf{k}} = -\frac{1}{N}\sum_{\mathbf{k}\prime} V_{\mathbf{k}\mathbf{k}\prime} \langle b_{\mathbf{k}} \rangle$. At arbitrary temperature the expression is given by:

$$\Delta_{\mathbf{k}} = -\frac{1}{N}\sum_{\mathbf{k}\prime} V_{\mathbf{k}\mathbf{k}\prime} \frac{\Delta_{\mathbf{k}\prime}}{2E_{\mathbf{k}\prime}} \tanh\left(\frac{E_{\mathbf{k}}}{2k_B T}\right) \tag{2.3.5}$$

In the Ginzburg-Landau theory, when the temperature is close to T_c, the spatially dependent gap is locally proportional to the order parameter. BCS predicts a relation between the energy gap and the phonon energy:

$$\Delta_0 \cong 2\omega_D \exp\left(\frac{-1}{V_0 D(E_F)}\right) \tag{2.3.6}$$

with ω_D the Debye frequency. For weak coupling the energy of the gap is related to the superconducting transition temperature (T_c) by:

$$\frac{2\Delta_0}{k_B T_c} = 3.52 \tag{2.3.7}$$

2.4. Unconventional superconductivity

The BCS theory shows how the Fermi sea is unstable against the formation of Cooper pairs under a small positive interaction. In the conventional case the pairing is produced by phonon-electron interactions. However, in BCS the derivation of the gap equation 2.3.5 is more general. Consequence of the BCS gap equation is that a repulsive interaction can condensate Cooper pairs if the gaps $\Delta_{\mathbf{k}}$ and $\Delta_{\mathbf{k}\prime}$ have opposite signs. Therefore, $\Delta_{\mathbf{k}}$ can have non trivial symmetry, when the interaction are not phonon mediated [19].

Due to the proximity to a magnetically ordered phase of cuprates and Fe-based supercon-ductors (Fig. 2.4.1 (a-b)) spin fluctuations have been proposed as attractive interactions for Cooper pairing [21, 22]. The idea behind spin fluctuations is that in systems located close to a magnetic instability, the screened Coulomb interaction between fermions can be approximated by an effective interaction mediated by collective fluctuations in the spin channel.

The cuprates (the family of superconducting compounds based on Cu-O [23, 24]) are commonly used as example of spin-fluctuations mediated pair interaction [20]. As shown

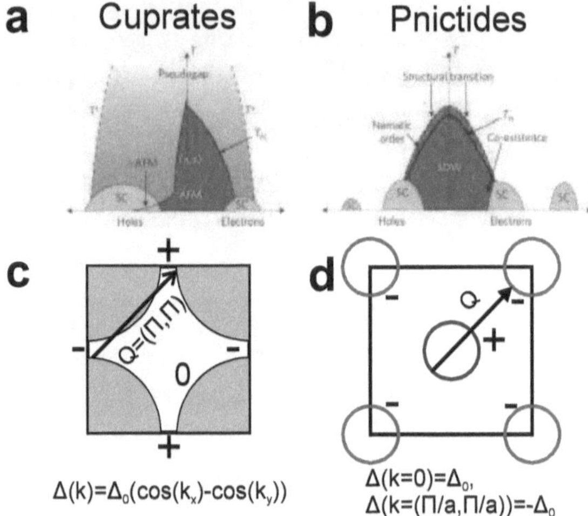

Figure 2.4.1.: (a-b) Schematic phase diagram of cuprates and Fe-based superconductors for hole- and electron doping. (c-d) Fermi surface of Cuprates and Fe-based superconductors with spin-fluctuations Q vector proposed as pairing mechanism with $d-$wave symmetry for cuprates and $s-$wave (s^{+-}) for Iron-based superconductors. Figure adapted from Refs. [20, 21]

in Fig. 2.4.1 (c) $\Delta_{\mathbf{k}}$ at the Fermi surface has different sign between points separated by the vector $Q = (\pi/2,\pi/2)$. This gap symmetry with the gap changing sign in the diagonals is called $d_{x^2+y^2}$ symmetry. In Fe-based superconductors the superconducting gap function is an open question under investigation [25–27]. In these materials, the Fermi surface is more complex. The sketch in Fig. 2.4.1 (d)) has a hole pocket in the centre (Γ-point) and electrons pockets in the corner of the two-Fe unit cell. The hole and electron pockets are well nested with vector $Q_1 = (\pi/a,0)$ and $Q_2 = (0,\pi/a)$. This Fermi surface favours a gap function Δ_k that changes sign between k and $k + Q_{1,2}$. The solution that does not lower the symmetry of the lattice is a s-wave symmetry, often changing gap sign from electron to hole pocket and address as $s_{\pm}$ [21]. An important distinction between $d-$wave and $s-$wave gap symmetry is that a $d-$wave gap has zeroes in the Fermi surface. This implies that the quasiparticle density of states for $d-$symmetry is non-zero in some points of the Fermi surface. In Fe-based superconductors the node is avoided due to the multiorbital nature of the Fermi surface.

2.5. Summary

For correlated electron systems the complexity arises from the competing degrees of freedom: charge, lattice, orbit and spin. The balance between these different degrees of freedom determines the ground state. This balance can be effortlessly disrupted by instabilities, which promote new ground states or complex collective behaviour. Furthermore, fluctuations play an important role for possible phase transitions. For instance in cuprates and iron-based superconductors, magnetic (spin) fluctuations are widely believed to be responsible for superconductivity. However, both material possess complicated phase diagrams populated by several different ordered phases [2, 9, 11, 28–35]. Consequently determining the driving force behind these phases is a complex task. This **book** tries to contribute to the physics of correlated electron systems in two different ways:

- Probing the electronic structure of a weakly doped AFM insulator in Sr_2IrO_4.

- To determine elusive electronic order in iron-based superconductors (in particular Li doped NaFeAs).

3. Experimental methods

Scanning Probe Microscopy (SPM) is a family of surface probing methods. Here, a local probe (typically a metallic tip) is moved over a sample surface yielding data about its morphology (see Fig. 3.0.1). Parable of this measurement mechanism is a person finger running across words written in Braille. The finger acts as the local probe in the SPM, and the letters represent the surface of interest. An outstanding characteristic of all SPMs is the high spatial resolution of the measurements, which is only limited by the size of the probe. Its displacement is controlled by piezoelectric motors whose accuracy is smaller than the probe size (down to one atom in the ideal case, $\leq 1\,\text{Å}$) (see the example sketched in Fig. 3.1.1). In contrast to optical microscopy, diffraction does not limit the astonishing resolution of SPMs. However, the resolution varies from technique to technique depending on the probe-sample interaction. A common denominator of this family of probes is that the data is obtained in 2D arrays and pictured in a false colour computer image. This idea is applied to a vast number of techniques such as scanning tunnelling microscopy (STM); atomic force microscopy (AFM); magnetic force microscopy (MFM); scanning near-field optical microscopy (SNOM)...

3.1. Scanning tunnelling microscopy

STM was first developed in 1981 by Gerd Binnig and Heinrich Rohrer [36]. The technique is based on the quantum tunnelling effect, one of the fundamental features of quantum mechanics, where the electronic wave function is allowed to propagate through a classically forbidden barrier. The phenomenon was observed first in planar junctions and used to measure the electronic structure of superconductors [37].

STM uses the vacuum between the tip and sample as a tunnelling barrier between the electron. It uses a closed-circuit configuration, where an external voltage provides energy to the electron in tip or sample to tunnel through the barrier, and therefore, to create a tunnelling current. In the scanning mode, the tip moves over the surface of the sample. While scanning, the tunnelling current is kept constant by an electrical feedback loop that varies the distance between the tip and sample. The changes in the tip height (z-direction) are recorded as a function of x and y and plotted in a 2D topography image.

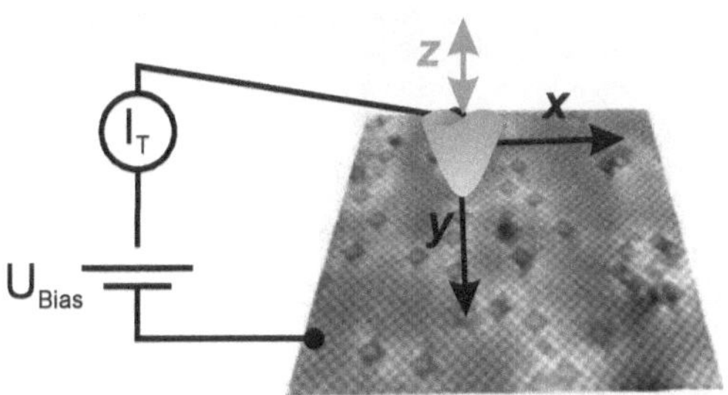

Figure 3.0.1.: Schematic drawing of a tip scanning a sample surface in the xy plane with tunnelling current I_T and bias voltage U_{Bias}. Piezoelectric actuators vary the position of the tip (x,y).

3.1.1. Quantum tunnelling

The tunnelling equations can be solved in the simplest model considering tip and sample as ideal metals. A vacuum gap z separates the two of them. A bias voltage U_B is applied to tip or sample, shifting the Fermi energies by eU_B with respect to the other. Here, it is considered the case where a positive voltage increases the tip energy. The two work functions, Φ_{sample} and Φ_{tip} plus eU_B, separated by z, form a trapezoidal barrier for the electrons (as sketched in Fig. 3.1.1).

For a tip electron, the finite probability of being localized in the sample is given by:

$$|\psi(z)|^2 = |\psi(0)|^2 e^{-2\kappa z} \tag{3.1.1}$$

where κ is the decay constant in the vacuum region:

$$\kappa = \sqrt{\frac{m_0}{\hbar^2}(\Phi_{tip} + \Phi_{sample} - eU_B)} \tag{3.1.2}$$

m_0 is the free electron mass. Ref. [38] gives a detailed quantum mechanical derivation.

Tunnelling current

The tunnelling current from tip to sample can be calculated using the *local density of states* (LDOS), i.e. the density of states found at a specific location per energy interval:

$$\rho(\vec{r},E) = \sum_v |\psi_v(\vec{r})|^2 \delta(E_v - E) \tag{3.1.3}$$

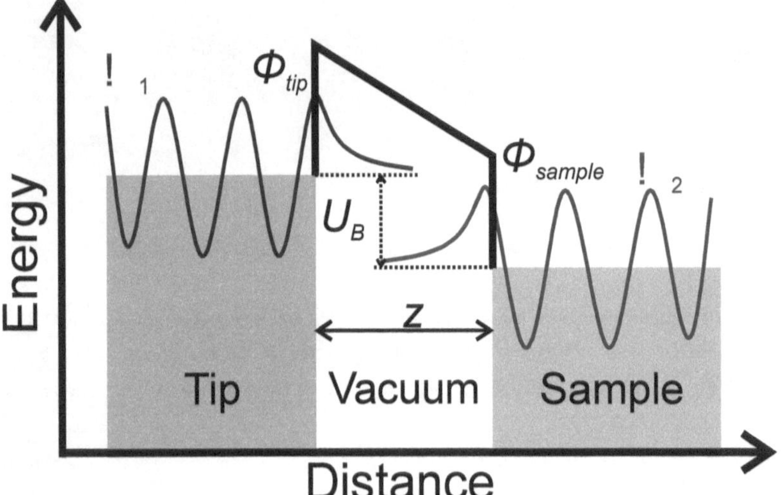

Figure 3.1.1.: Schematic view of the tunnelling process between a metallic tip (right-hand side) and a sample (left-hand side). When a bias voltage (U_B) is applied to the sample, the electrons in the valence band of the tip are able to tunnel into unoccupied states of the sample. The tunnelling barrier shows a trapezoidal-like shape due to the difference of the tip work function (Φ_{tip}) with respect to bias voltage, plus the sample work function (Φ_{sample}).

Now the tunnelling current is :

$$I_{t \longrightarrow s} = \frac{4\pi e}{\hbar} \int_{-\infty}^{\infty} \rho_t(\epsilon - eU_B)\rho_s(\epsilon) f_t(\epsilon - eU_B)(1 - f_s(\epsilon))|M(\epsilon - eU_B,\epsilon)|^2 d\epsilon \qquad (3.1.4)$$

Here, the energies are relative to the Fermi energy of the sample and tip. $f(\epsilon)$ is the Fermi-Dirac temperature distribution for electrons, $M(\epsilon_t,\epsilon_s)$ the tunnelling matrix element. In a one dimensional simplification, it is given by:

$$|M(\epsilon - eU_B,\epsilon)|^2 = \exp\left[-2z\sqrt{\frac{m_e}{\hbar^2}}(\Phi_{tip} + \Phi_{sample} - eU_B + 2\epsilon)\right] \qquad (3.1.5)$$

Only occupied states of the tip and unoccupied states of the sample are counted due to the Fermi-Dirac distributions. Furthermore, $I_{t \longrightarrow s}$ depends linearly in the LDOS of tip and sample. For the total tunnelling current I_T, the contributions in both directions are taken into account, and the result is:

$$I_T = \frac{4\pi e}{\hbar} \int_{-\infty}^{\infty} \rho_t(\epsilon - eU_B)\rho_s(\epsilon)\left(f_t(\epsilon - eU_B) - f_s(\epsilon)\right)|M(\epsilon - eU_B,\epsilon)|^2 d\epsilon \qquad (3.1.6)$$

The equation 3.1.6 gives an easy understanding of the tunnelling current. However, for real applications, a determination of the tunnelling matrix element in a more realistic approximation is needed. In 1961 Bardeen used the first-order time-dependent perturbation theory to compute a realistic expression of the tunnelling matrix element [39]:

$$M_{ts} = \frac{2\pi}{\hbar} \int d\vec{S} \cdot (\psi_t^\dagger \vec{\nabla} \psi_s - \psi_s \vec{\nabla} \psi_t^\dagger) \qquad (3.1.7)$$

With the integral over any surface lying entirely within the vacuum region. Eq. 3.1.7 shows how to calculate the tunnelling matrix element. However, the calculation is only possible if the wave functions from tip and sample are known.

For most experimental conditions the tip atomic structure is unknown, and therefore its wave function cannot be determined. Tersoff and Hamann gave a solution to the uncertainty of the tip wave function in first-order perturbation theory [40]. They considered an atomically sharp tip and assumed that only the atom that is closest to the sample contributes to the tunnelling process. The wave function for this atom was considered as a spherical s-like orbital and the density of states constant in the energy interval under consideration. Only elastic tunnelling processes were taken into account:

$$M_s = \frac{2\pi C\hbar^2}{\kappa m_e}\psi(\vec{r}) \qquad (3.1.8)$$

With C constant, in the Tersoff-Hamann solution, the tunnelling matrix does not depend on the tip wavefunction but only in the outer atom of the tip at $\vec{r}$. The calculated expression in the low voltage approximation ($eU_B << \Phi$) for the tunnelling current is:

$$I_T = \frac{16\pi^3 C^2 \hbar^3 e}{\kappa^2 m_e^2} \rho_t \int_0^{eU_B} \rho_s(\epsilon)d\epsilon \qquad (3.1.9)$$

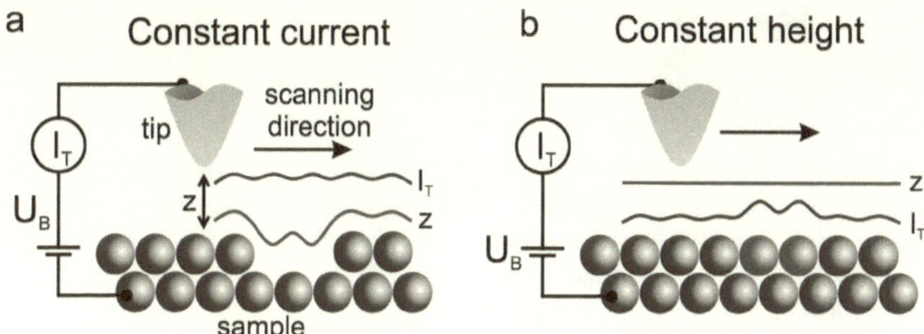

Figure 3.1.2.: STM most common imaging modes. (a) Sketch of a constant current topography measurement. When the tip is moving in the scanning direction z is adjusted, so the tunnelling current (I_T) is kept constant. (b) Sketch of the constant height mode. In this case, z is kept constant, and the variation of I_T is recorded.

3.1.2. Imaging

STM exploits the exponential relation of the tunnelling current (I_T) with the tip-sample distance (z). The tip scans the sample surface and acquires the so-called topography images. There are two main work modes (see Fig. 3.1.2). On the one hand, the constant current mode, which is the most extended. The tip scans the sample with a fixed I_T (set-point) line by line, adjusting z with a feedback loop, which drives the tip up and down to keep the tunnelling current equal to the set point. The variation of z (with its position recorded in an (x,y) array) reveals the surface topography. Since the tip height is adjusted at every point, this is the safest mode for scanning surfaces with unknown roughness. On the other hand, the other scanning mode is the constant height mode. Now, z is kept fix while scanning and recording the variation of I_T. In this mode, the $I_T(x,y)$ correspondent image does reveal a convolution of the surface roughness and the LDOS. It allows faster scanning than constant height since the feedback loop is not needed.

3.1.3. Scanning tunnelling spectroscopy

At the heart of the STM is the capability to acquire local spectroscopy data with atomic resolution. Equation 3.1.9 showed the connection between the tunnelling current and the LDOS of the sample. Scanning tunnelling spectroscopy (STS) benefits from this relation measuring the LDOS of the sample through the derivative of the tunnelling conductance.

$$G(\mathbf{r}, U_B) \equiv \left(\frac{dI_T}{dU_B}(\mathbf{r}, U_B) \right) \propto \rho_s(\mathbf{r}, eU_B) \qquad (3.1.10)$$

Point spectroscopy

A point spectrum is recorded at a particular location on the sample surface (x,y). For this spectrum, the changes in I_T are recorded during the sweep of the bias voltage U_B. The tip-sample distance is kept constant during the sweep (The feedback loop is switched off). The result is an I-V curve. The derivative of such an I-V curve is proportional to the LDOS of the sample. While it is possible to execute a numerical derivation of the curve, it usually introduces noise in the measurements and therefore requires smoothing the resulting curve. Consequently, direct methods are employed to measure the dI/dU spectra.

The lock-in technique provides spectroscopic data with high signal to noise ratio. In this technique, a small sinusoidal voltage ($V_m \cos(\omega_m \tau)$) is used to modulate the tunnelling voltage. The frequency of the modulation is set to much higher values than the regulating speed of the feedback loop. This difference minimises the influence of the modulation in the acquired topography images.

Now the modulated tunnelling current can be written as:

$$I_T = I_T(U + V_m \cos(\omega_m \tau)) \tag{3.1.11}$$

Using the Taylor expansion:

$$I(U + V_m \cos(\omega_m \tau)) = I(U) + a_0 \frac{dI(U)}{dU} V_m \cos(\omega_m \tau) + a_1 \frac{d^2 I(U)}{dU^2} V_m^2 \cos(2\omega_m \tau) + ... \tag{3.1.12}$$

Where a_n are the Taylor coefficients. The first harmonic of the tunnelling current Taylor expansion is proportional to the tunnelling differential conductance. In the Lock-in technique, the higher harmonic terms are normally suppressed with a low-pass filter. Finally, the response signal from the current channel is demodulated. This signal is proportional to the LDOS of the sample.

For STS, there are two main limitations of energy resolution. The temperature-dependent energy broadening of the Fermi-Dirac distribution and the blurring caused by the modulation voltage V_m. The Fermi -Dirac distribution is given by:

$$f_F(\epsilon) = \frac{1}{1 + e^{\epsilon/k_B T}} \tag{3.1.13}$$

With k_B the Boltzmann constant. The overall broadening is calculated through the full width at half maximum (FWHM) of f_F derivative.

$$\Delta E = \sqrt{(3.5 k_B T)^2 + (2.5 e V_{mod})^2} \tag{3.1.14}$$

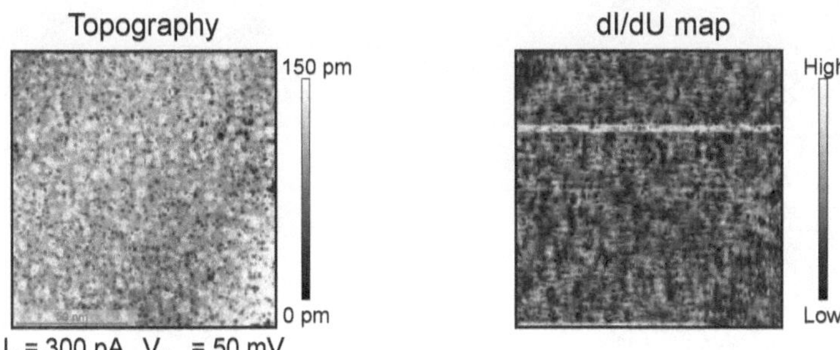

Figure 3.1.3.: 120 nm × 120 nm topographic and dI/dU map of the atomically re-solved surface of Na$_{0.97}$Li$_{0.03}$FeAs measured at $T = 5.2$ K. The data were recorded with bias voltage $U_{\text{bias}} = 50$ mV and tunnelling current $I_T = 300$ pA. The tunnelling conductance was measured with a lock-in technique with U_{mod}=1.6 mV. The topography shows a high abundance of crystalline defects. The dI/dU maps reveal the existence of electronic ordered patches along with the vertical or horizontal directions.

dI/dU map

The lock-in technique can register the changes of the tunnelling current during a constant current mode. The derivative $\frac{dI}{dU}(x,y)$ is acquired simultaneously to $I(x,y)$. Since the tip-sample distance is modified during the experiment, the data obtained is not truly proportional to LDOS. Nonetheless, it can help to visualise electronic changes in the sample surface at a specific U_B, especially if the surface roughness is negligible.

3.1.4. Spectroscopy-imaging STS

STS achieves its full potential with spectroscopy-imaging STS (SI-STS) [41]. It is the natural extension of the spectroscopy mode. A single $\frac{dI}{dU}$ spectrum is measured at every pixel in a chosen spatial grid. The differential conductance data set has three dimensions $g = \frac{dI}{dU}(x,y,U)$. Two of which correspond to spatial dimensions and the third is the specific energy. A common representation is in a false colour plot similar to topography for every energy value (see Fig. 3.1.4). The representation permits a visualization of the spatial dependence of the LDOS. When high energy resolution is measured with high spatial resolution, the measurement takes long times. Depending on the requirements, it varies from a few hours to several days. Typically the measurements are performed under low-temperatures to provide stability to the system and avoid thermal drift (displacement of the tip due to changes in the piezoelectric actuator).

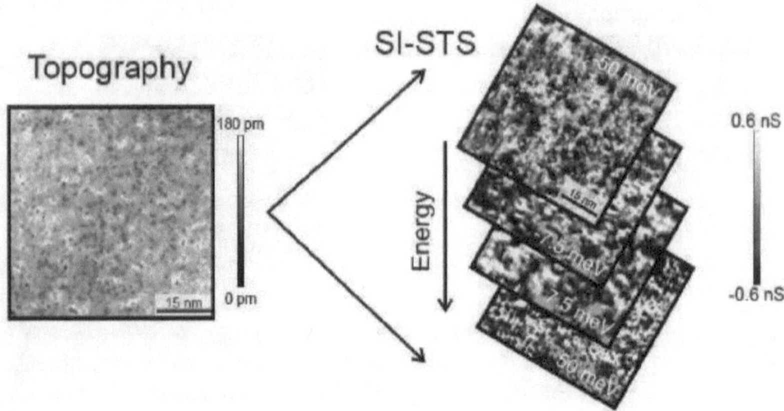

Figure 3.1.4.: Example of SI-STS for $Na_{0.96}Li_{0.04}FeAs$ measured at $T = 4.8$ K in a 60 nm × 60 nm area. The data was recorded with bias voltage $U_{bias} = 50$ mV and tunnelling current $I_T = 300$ pA. The correspondent topography is shown on the left. At the right, a set of conductance maps with energies between 50 to -50 meV are shown.

3.2. Quasiparticle interference

In an ideal metal surface free of defect, the electronic wave functions have k-space eigenstates and no spatial dependence.

$$LDOS(\mathbf{r},U) \propto \sum_k |\psi_k(\mathbf{r})|^2 \delta(eU - \epsilon(\mathbf{k})) \tag{3.2.1}$$

The scenario is changed when some source of disorder is present (impurities or crystal defects are the most common ones). The electron scattering with the disorder source mixes eigenstates with different k but are located at the same constant energy contour in k-space. The mixing of states produces a quasiparticle wave function with wave vector $\mathbf{q} = \mathbf{k}_2 - \mathbf{k}_1$. Early STM experiments showed wavelike patterns for noble-metals in the vicinity of steps and adatoms [42–44]

3.2.1. Fourier-transform STM

The quasiparticle interference (QPI) modulates the LDOS with wavelength $\lambda = 2\pi/|\mathbf{q}|$. SI-STM is a tool to measure such oscillation created by scattered electrons. These oscillations reflect the shape of the Fermi contours (fermiology). A method to determine the wave vector is the Fourier transform. Fourier-transform STM (FT-STM) provides a fast and

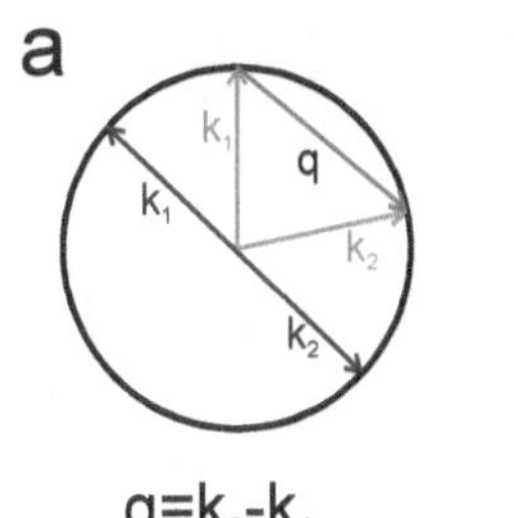

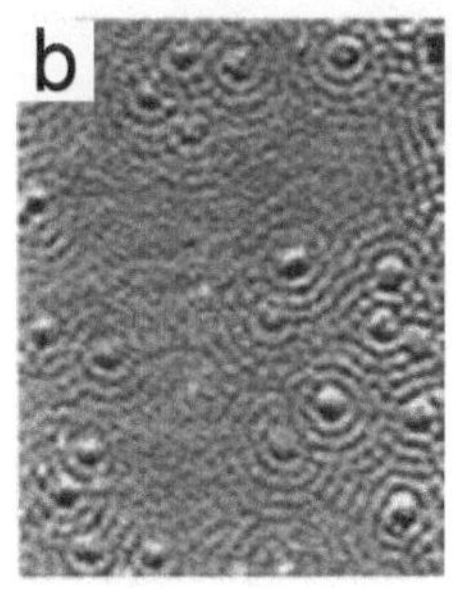

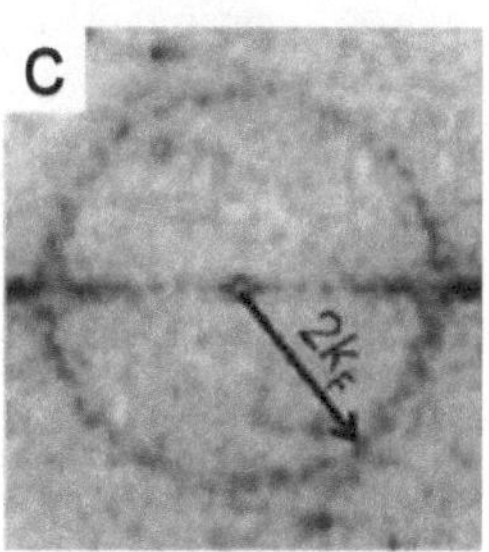

Figure 3.2.1.: (a) Constant energy Fermi contour. The scattering of an incident wave vector $\mathbf{k}_1$ is transferred to $\mathbf{k}_2$ after being scattered. Two cases are sketched: Backscattering is plotted in blue, with $\mathbf{q} = 2\mathbf{k}$. In orange is plotted the scattering of $\mathbf{k}_2$ with a random angle with respect to $\mathbf{k}_1$. (b and c) Show the example of a constant contour extracted from Friedel oscillations in Cu (111). (b) Constant current STM image of Cu (111). The Friedel oscillations are visible as ring-like waves around defects. (c) 2D Fourier transform of (b). The radius value corresponds to two times the Fermi wave vector measured on ARPES for a circular surface state. Figure adapted from Ref. [45]

accurate technique to determine two dimensional Fermi contours. This technique was successfully applied in metals to measure the Friedel oscillation [45, 46] (see Fig. 3.2.1) and later used in cuprate or IBS to measure the Bogoliubov-quasiparticles that scatter from weak impurities [47–50]. FT-STM has also shown pieces of evidence of C_4 symmetry breaking (nematicity) [30, 33, 51] and charge ordering [11].

Fourier transform of a conductance map in real space allows to obtain information from the reciprocal space ($\mathbf{k}$-space):

$$LDOS(\mathbf{k},U) = \rho(\mathbf{k},U) \propto -\Im\left(Tr\left[G(\mathbf{k},U)\right]\right) \tag{3.2.2}$$

QPI requires to compute the Green's function in the presence of impurities. A successful method is the T-matrix approximation [52–54]. To calculate the Fourier transformed local density of states the one-particle Green's function $G_0(\mathbf{k},U)$ is taken, plus a local potential scatterer with an assumed potential $V_{\mathbf{k},\mathbf{k}'} = V_{imp}$ in momentum space.

$$\rho(\mathbf{q},U) = \frac{1}{\pi}\sum_{\mathbf{k}}\Im G(\mathbf{k},\mathbf{k}\text{-}\mathbf{q},U) \tag{3.2.3}$$

Here, $G(\mathbf{k},\mathbf{k}',U)$ is the retarded Green's function in the presence of one single impurity and is related to the retarded Green's function of the bulk material $G(\mathbf{k},E)$ via

$$G(\mathbf{k}, \mathbf{k'},U) = G_0(\mathbf{k},U) + G_0(\mathbf{k},U)T_{\mathbf{k},\mathbf{k'}}(U)G_0(\mathbf{k'},U) \qquad (3.2.4)$$

Where $T_{\mathbf{k},\mathbf{k'}}(E)$ is the energy dependent t-matrix.

QPI analysis provides local information about the electronic structure above and below the Fermi level. However, it is not a direct measurement of the momentum dispersion in materials and comparison with direct techniques like ARPES and band structure calculation is required to understand the data.

3.3. Experimental set-up

All the experimental data in this **book** were collected using a home-built STM [55] called "Dip-stick STM". The Dip-stick has a tube design (Fig. 3.3.1) that allows to insert it on commercially available bath cryostat. The cryogenic vacuum is realized inside the tube. The whole system is lifted and damped to isolate from acoustic noise. The system has enormous stability that provides long measurement times at base temperature (up to 8 weeks using a homemade dewar of 200 litres of He^4). Furthermore, the system is well suited to measure at elevated temperatures. Therefore, it is the perfect tool to investigate electronic phases across phase diagrams.

The STM head is mounted at the end of the rod. It is designed in a Pan style [56, 57]. A piezo-electrical scanner tube is planted inside a sapphire prism which is a vertically driven (z-axis) by six piezoelectric walker stacks motors. The samples are cleaved, to obtain new and clean surface, inside the STM at base temperature, which is a significant advantage of the system. Air sensitive samples were mounted inside an Argon glove-box where the dip-stick is previously introduced.

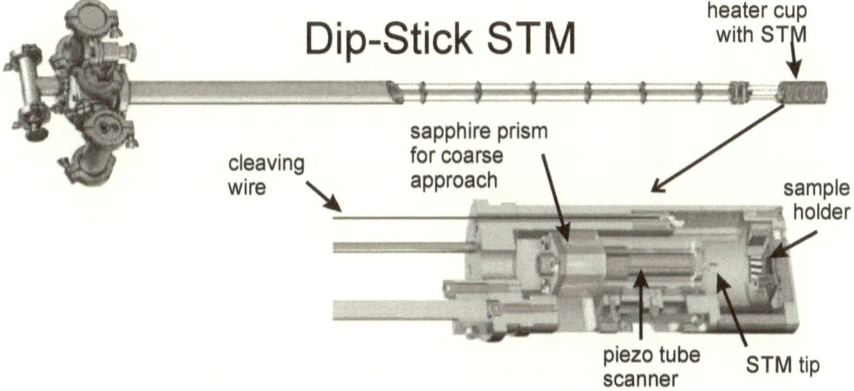

Figure 3.3.1.: Design of the Dip- Stick STM system with a section through the stainless steel tube. The image magnification shows the inside of the STM head, which allows in-situ cleaving of the samples. Figure adapted from [55]

4. Iridates

4.1. Introduction

The family of iridium oxides (iridates) exhibit a unique competition between Coulomb interactions, crystal fields splitting and strong spin-orbit coupling (SOC) [58, 59]. A general energy diagram of the configuration of the $5d$ orbitals in the iridates is plotted in Fig 4.1.1 (a-e). First, the states are split by the crystal field, which is usually octahedral. Under this symmetry, the $5d$ states are divided into a t_{2g} triplet and e_g doublet by a crystal field energy 10Dq. For a $5d$ transition metal oxide, the energy distance to the doublet is sufficiently large to neglect the e_g sates. Moreover, the Iridium-oxide or iridates gained attention due to their structural and electronic similitude with the cuprate superconductors.

For Ir ($Z = 77$) the relativistic SOC is strong enough to split the t_{2g} multiplet into an effective $J_{eff} = 1/2$ singlet and $J_{eff} = 3/2$ quartets. The J_{eff} states form very narrow bands (even small correlations energy leads to a $J_{eff} = 1/2$ Mott ground state [58]) with unique electronic and magnetic behaviours [59]. This fascinating interplay of electronic interactions generates numerous novel quantum phases which include unconventional magnetism, spin liquids, and strongly correlated topological phases [59–63].

4.2. Sr_2IrO_4

Sr_2IrO_4 is the standard-bearer of the iridates novel physics. It is the first known spin-orbit assisted $J_{eff} = 1/2$ Mott insulator [58, 64] with a Neel temperature $T_N = 240\ K$ [65, 66]. The crystal structure of Sr_2IrO_4 is that of a layered perovskite consisting of quasi-2D planes of IrO_6 octahedra and sandwiched Sr sheets [65–68] as shown in Fig 4.1.1 (f). An important structural feature of the ground state is a rotation of the octahedra of $11°$ with respect to the c-axis. This rotation corresponds to an in-plane distortion of the Ir_1-O-Ir_2 bond angle. The electronic and magnetic properties are very sensitive to the rotation of the octahedra [59]. For example, the antiferromagnetic moments are canted $13°$ away from the a-axis tracking the rotation of the IrO_6 [69–71].

Sr_2IrO_4 is isostructural to La_2CuO_4, a prominent parent compound of the cuprates. The electronic and magnetic similarities between Sr_2IrO_4 and La_2CuO_4 (like quasi-2D magnon

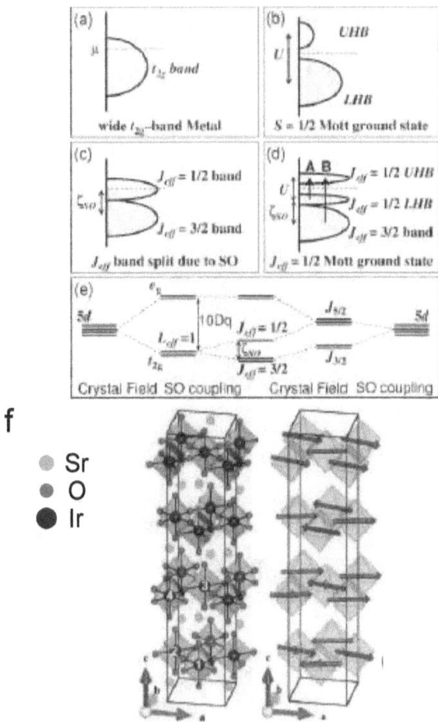

Figure 4.1.1.: Schematic energy diagrams for the 5d configuration (a) without SOC and U, (b) with an unrealistically large U but no SOC, (c) with SOC but no U, and (d) with SO and U. Possible optical transitions A and B are indicated by arrows. (e) $5d$ level splitting by the crystal field and SO coupling. (f) Crystal structure of Sr_2IrO_4 and sketch of the magnetic moments orientations. Figure adapted from Ref. [58]

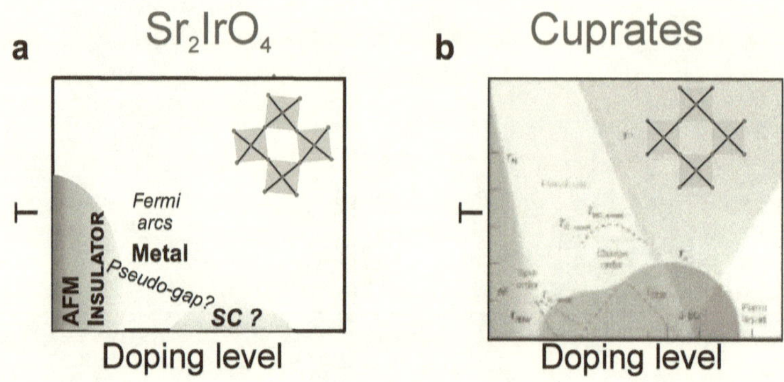

Figure 4.2.1.: Comparison between the sketched phase diagram of the Sr_2IrO_4 with results taken from Refs. [72–80] and a general phase diagram of the cuprates adapted from Ref. [81].

dispersion [64, 82]) led to the proposal of superconductivity [73]. Although there is no evidence of superconductivity yet, it is interesting to review the experimental attempts to drive the system away from the Mott insulating state through pressure, electron- or hole-doping. It is expected that the insulating states collapse upon pressure, and the system reaches a metallic state. That has not been the case for Sr_2IrO_4 or other iridates, where a metallic sate has not been found up to 40 GPa [59, 83, 84] which remarks the novelty of the magnetic and insulating ground state. Electron- or hole-doping can be achieved by chemical substitution. Electron- or hole-doping-doping are achieved via doping in the Sr site, with La- [74–77] introducing extra electrons and K doping introducing extra holes [74]. The Ir site can be doped with Ru for electron-doping [85, 86] or isoelectronic Rh [87, 88]. Doping can also be achieved by the introduction of oxygen deficiencies [72]. In contrast with the pressure studied all the different chemical substitution turn the system into a metallic state. The metallic state is not all plain and featureless. In La doping there are observations of a pseudogap (a weakly conducting state with an anisotropic energy gap [89–91]) in the Fermi arcs [76], and an STS study in real space had shown that the Mott gap collapses at 4% and form puddles of the pseudogap [80] (Fig. 4.2.2 (e)). A comparison of the phase diagram of Sr_2IrO_4 and the cuprates general phase diagram is shown in Fig. 4.2.1. The most encouraging results regarding a superconducting phase come from studies of in-situ electron doping via K coverage of the surface. Upon doping with alkali disconnected segments of zero-energy states (Fermi arcs) have been observed by ARPES [92] (Fig. 4.2.2 (d)). The same experiment has reported signs of a d-wave

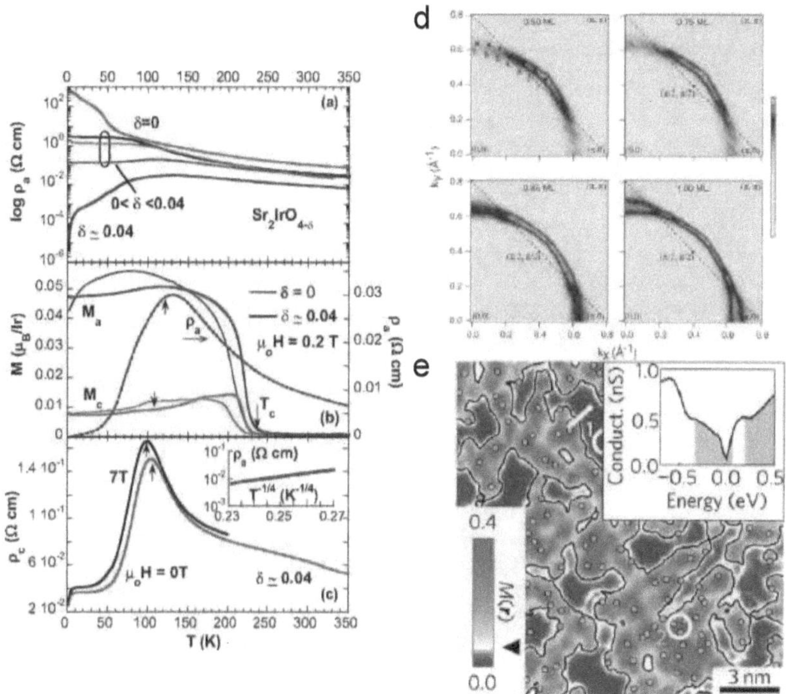

Figure 4.2.2.: (a) Temperature dependence of ρ_a for several values of δ in $Sr_2IrO_{4-\delta}$ (b) $a-$ and $c-$ axis magnetization. (c) ρ_c at 0 and 7 T. The inset $\log \rho_a$ vs $T^{-1/4}$. (d) Constant-energy intensity maps taken at the Fermi energy for surface coverage of 0.5 ML, 0.75 ML, 0.85 ML, and 1 ML. By varying the surface coverage of potassium atoms deposited on the surface of Sr_2IrO_4, the electronic structure evolves across the entire phase diagram from a Mott insulating state to a normal metallic state, via a strange metallic phase. (e) Phase-separated Mott/pseudogap electronic structure of $(Sr_{0.945}La_{0.055})_2IrO_4$ the tunnelling conductance (insets) for the areas in red resembles pseudogap in cuprates. Fig adapted from Refs. [72, 80, 92]

symmetric gap at the Fermi arcs [79]. These are commons observations in the cuprates superconductors [9, 11]. Similarly, a d-wave gap spectral feature (nodeless-gap) has been reported by STS [78]. All these are surface-doping studies and the differences with the La bulk doping are still unknown.

The emergence of metallic transport in the spin-orbit assisted Mott insulator Sr$_2$IrO$_4$ offers new insides into the metal to insulator transition (MIT). The MIT in correlated electron systems is one of the open challenges in condensed matter physics [93, 94]. Near the transition point, several competing states are observed in the spin, charge and orbital degrees of freedom [2, 94]. In particular, Mott insulators show strong spatial inhomogeneity [95–98]. With this in mind, STM is an ideal tool to locally study the electronic structure of the Sr$_2$IrO$_4$ and disentangle the relation between Mott physics and magnetism.

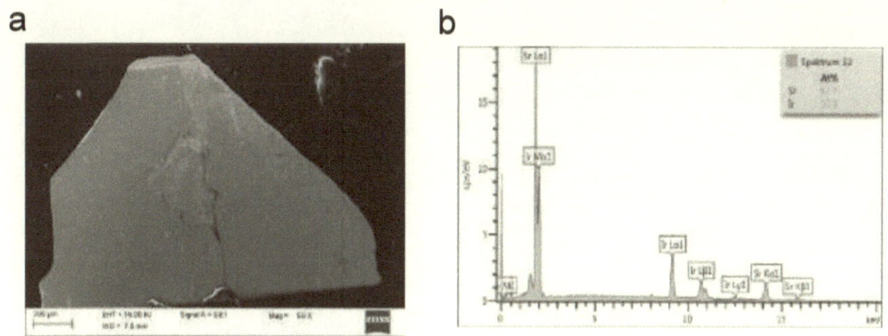

Figure 4.3.1.: (a) SEM image of as-growth Sr_2IrO_4 single crystal. (b) EDX composition determination. The sample shows a small Sr deficiency.

4.3. Experimental data

4.3.1. Sr_2IrO_4 single crystals

The Sr_2IrO_4 single crystals used in these experiments were grown in IFW Dresden (by K. Manna, A. Maljuk and S. Wurmehl) with the flux method [66, 99]. The growth was done inside a Pt crucible where the $SrCO_3$ and IrO_2 powders components (with 4N purity) were mixed with $SrCl_2$ flux with a 1:5 sample-to-flux weight ratio. After the temperature ramp, the mixture was heated to 1210 °C for twelve hours and then slowly cooled to 1000 °C with a cooling rate of 4 °C/h. Which was followed by a rapid cool-down to room temperature at 150 °C/h. the flux is dissolved in water and crystals are filtered.

A standard 4-probe technique (5 K - 300 K) was used in the as-grown $Sr_2IrO_{4-\delta}$ single crystals to measure the in-plane resistance to control the insulating behaviour of the samples. Fig. 4.3.2 shows representative resistivity data for different samples. The resistivity shows a semiconductor-like temperature dependence (Fig. 4.3.2), which proves that the sample is still close to the Mott insulator regime. The STM study was performed on the sample with reduced resistivity, which allowed to perform high-resolution STM/STS measurements even at very low temperature ($T < 10$ K) for the first time. The STM topographies shown here were obtained after cold-cleaving the sample crystal at low temperatures in the cryogenic vacuum.

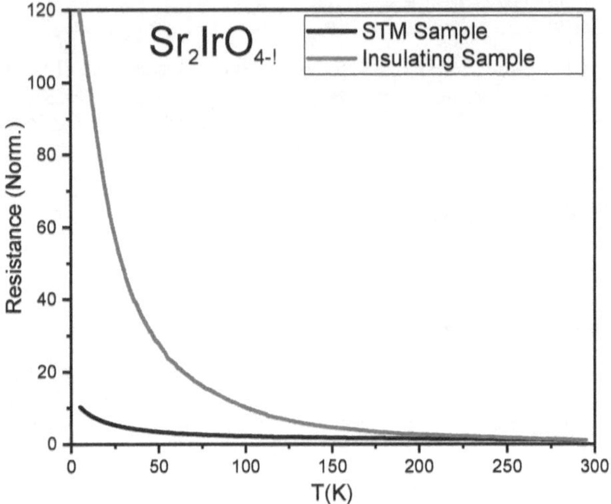

Figure 4.3.2.: Normalized in-plane resistivity of selected $Sr_2IrO_{4-\delta}$ samples. The strong reduction of the low-temperature upturn of the resistivity of the sample labelled 'STM Sample' evidences a significant amount of oxygen vacancies. Transport data measured by J. Schoop

4.3.2. Topography

This section describes the measured topographies and recognizes the main crystalline defects seen on the sample surface. It is important to note that measuring STM/STS on stoichiometric Sr_2IrO_4 at cryogenic temperatures is challenging because the samples usually become too insulating. In that respect, I. Battisti et al. [100] investigated the effect of bad screening in the STS measurements. They studied how the electric field generated by the tip can partially penetrate the sample surface and is seen as circular charge areas in the conductance maps. The measured gap can be influenced by tip-induced band bending. Other tunnelling experiments can be found in the Refs. [75, 78, 80, 100–102]. Those challenges were not present in this work due to the poor resistivity of the as-grown single crystal described in Sec. 4.3.1, which allowed good tunnelling conditions.

Different atomically resolved flat surfaces are displayed in Fig. 4.3.3. With areas of 20 nm $\times$ 20 nm (a) and 14 nm $\times$ 14 nm (b and c). The easy cleavage plane of Sr_2IrO_4 lays in between the SrO layers, as shown in Fig. 4.3.3(e). The Fourier transformation (Fig. 4.3.3(d)) of Fig. 4.3.3(b) presents square the Bragg peaks, which correspond to the first Sr atoms. The corresponding reciprocal lattice vector are marked as $Q_x = (1,0)$ and $Q_y = (0,1)$ Fig. 4.3.3 (e)). Additionally, the vectors $Q_a = (-1/2,1/2)$ and $Q_b = (1/2,1/2)$ are observed. They correspond to the 1x1 superstructure produced by the rotation of the IrO_6 octahedra.

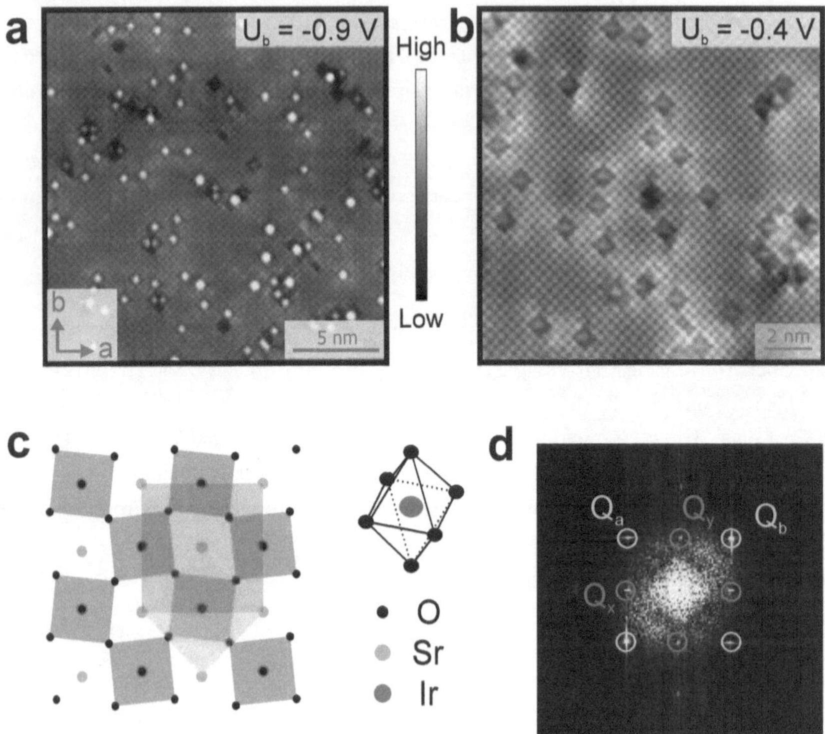

Figure 4.3.3.: Topographic images of atomically resolved clean surfaces measured at $T = 8.8$ K. (a) 20 nm × 20 nm topography of the sample surface at bias voltage $U_{\text{bias}} = 1.0$ V and tunnelling current $I_T = 200$ pA. (b) 14 nm × 14 nm topography of the sample surface in a different location from (a) measured at with tunnelling current $I_T = 200$ pA and $U_{\text{bias}} = -0.4$ V. (d) shows the Fourier transform of (b) with $Q_{x,y}$ vector of Sr lattice (red) and $Q_{a,b}$ for the 1x1 superstructure (blue). (c) depicts a sketch of the top view of crystalline structure. The Sr lattice cell is marked with a red square. The rotation of the IrO_6 octahedra produces a 1x1 superstructure marked by a blue square.

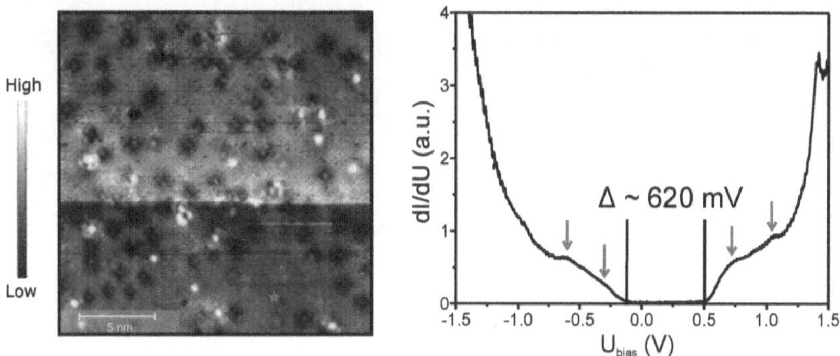

Figure 4.3.4.: Representative large scale tunnelling conductance spectrum taken on a clean place at $T = 8.8$ K. The corresponding topography image of 20.5 nm × 20.5 nm measured with $I_{Tunnel} = 200$ pA and $U_{bias} = 1$ V is plotted on the left. The spectrum (displayed on the right) was taken at the position marked in red. The dI/dU are measured with an external lock-in amplifier with $f_{mod} = 1.111$ kHz and $U_{mod} = 20$ mV (rms).

4.3.3. Spectroscopy on clean areas

Fig. 4.3.4 (right side) shows a representative tunnelling conductance (dI/dU) spectrum measured at $T = 8.8$ K at a place free of defects. The spectrum reveals a measurable gap of size $\Delta \approx 620$ meV where $dI/dU \approx 0$ between about -110 mV and 510 mV. This measurement is in agreement with other tunnelling experiments of stoichiometric samples measured at elevated temperatures [78, 101] and optical spectra, which show the first peak at 510 mV [58, 103]. For bias voltages below and above this gap, the tunnelling conductance reveals a shoulder-like increase which straightforwardly can be associated with the density of states of the lower and higher Hubbard bands. At $|U_{bias}| \gtrsim 1.3$ V, the dI/dU intensity increases vigorously, which can be attributed to further high-energy states, which have been observed with optical spectroscopy [104] above 1 eV, in qualitative agreement with the present data. Energy dependence of the tunnelling matrix element cannot be excluded. It could influence on high bias voltage measurement. Therefore, the interpretation of the tunnelling spectra is restricted to $U_{bias} \lesssim 1.2$ V. Furthermore, fine-structures of the peak- or shoulder-like anomalies can be distinguished in the tunnelling spectra close to the gap. They are signalled with red arrows in Fig. 4.3.4.

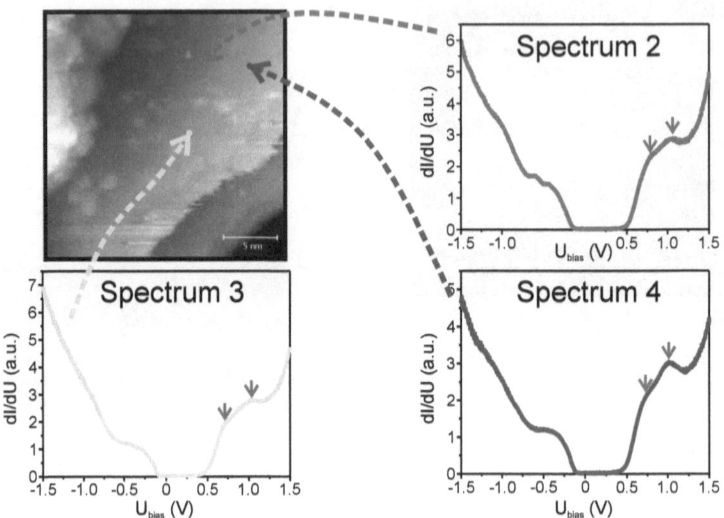

Figure 4.3.5.: Topography image of 22.5 nm × 22.5 nm showing a terrace of the crystal, measured at $T = 8.9$ K with $I_{Tunnel} = 200$ pA and $U_{bias} = 900$ mV, with spectra taken at different locations. The differential conductances are measured with large bias values (-1.5 V to 1.5 V) with an external lock-in amplifier with $f_{mod} = 1.111$ kHz and $U_{mod} = 20$ mV (rms).

Different spectra are shown in this section to demonstrate the representative character of the dI/dU spectrum shown in Fig 4.3.4. These spectra were taken at different points in different areas (shown in Figs. 4.3.5 and 4.3.6).

The topography in Figs. 4.3.5 presents a terrace between steps in the cleaved surface of Sr_2IrO_4. Even without the best experimental conditions, the dI/dU measured with large bias values (-1.5 V to 1.5 V) prove that the fine-structure features are still present.

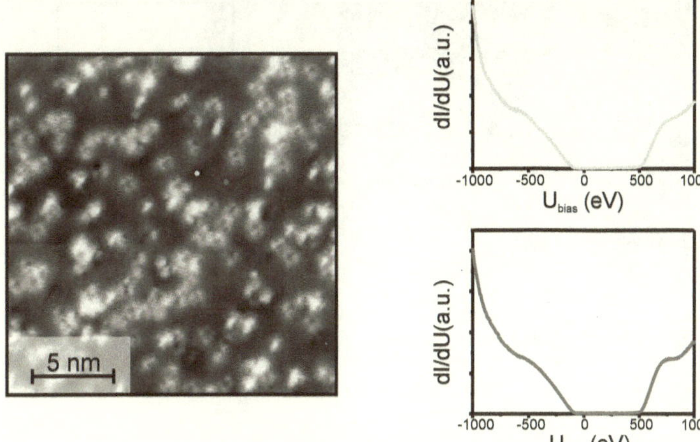

Figure 4.3.6.: Topography image of 20.5 nm × 20.5 nm measured at $T = 8.9$ K with $I_{Tunnel} = 200$ pA and $U_{bias} = 1$ V, with the spectra taken at two positions of the surface (marked in red and yellow). The dI/dU are measured with an external lock-in amplifier with $f_{mod} = 1.111$ kHz and $U_{mod} = 20$ mV (rms).

Another proof of the universality of fine-structures present in the spectra is measured by SI-STM (plotted in 4.3.7). Fig. 4.3.7(b) display the spectra taken along the yellow line inside a clean area of Fig. 4.3.7(a). All spectra show the first peak as well as the beginning of the second peak in a similar way.

The results here have proven the persistence of fine-structure peaks found beyond the LHB and UHB.

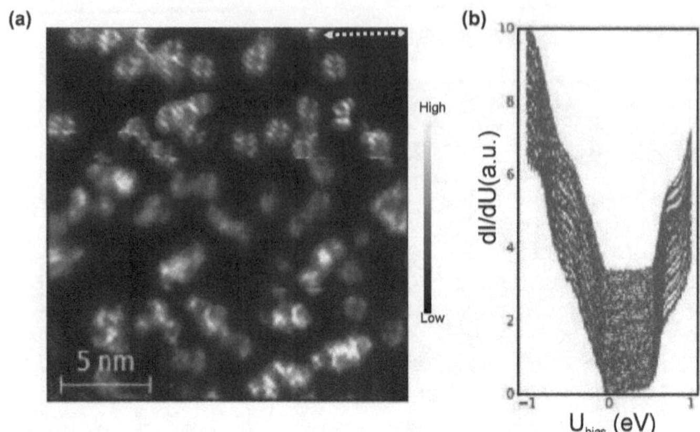

Figure 4.3.7.: (a)SI-STM map at 303.7 meV. Measured at $T = 9.3$ K with $I_{Tunnel} =$ 300 pA and $U_{bias} = 1.0$ V, the dI/dU spectra were measured with an external lock-in amplifier with $f_{mod} = 1.111$ kHz and $U_{mod} = 15$ mV (rms) for 271 energy point from -1.0 V to 1.0 V in an area of 20 nm $\times$ 20 nm with a 224×224 grid. (b) Local density of states spectra along the yellow line in (a).

4.4. Spin-polaron

The tunnelling conductance of Sr_2IrO_4 (see Fig. 4.3.4) possess fine-structures of the peak-
or shoulder-like anomalies. It is natural to look for signs of the magnetism in the origin of
these structures. Here a model based on spin-polaron excitations is used.

4.4.1. Spin-polaron model

The spin-polaron is a quasiparticle resulting in the coupling of a moving charge (electron or hole) with the elementary spin excitations of the antiferromagnetic lattice, which
have strong on-site Coulomb interactions. It touches one of the most fundamental open
problems in contemporary condensed matter physics as described in section 2.2.2. The
spin-polaron is characterized by an environment of misaligned spins (Fig. 4.4.1(a)) form-
ing an effective confinement potential, where the charge can occupy excited states of different orbital character [105] (see Fig. 4.4.1(b)). In the one-particle spectral function,
these excitations manifest themselves by the occurrence of a rather flat and ladder-like
structure (Fig. 4.4.1(c)). Due to quantum fluctuations, the spin defects can relax, and the
quasiparticle becomes dispersive.

This problem is of significant relevance for understanding high-temperature superconductivity [13], however clear experimental spectroscopic signatures of these internal degrees
of freedom are scarce.

The quasi-2D Sr_2IrO_4 has clear signatures of the AF Mott physics. Thanks to the
similarities with the cuprates, a multi-orbital 2D Hubbard model with spin-orbit coupling
was proposed as a model for the underlying electron system. As seen in section 2.2, the one-
band Hubbard model has an AFM ground state. This property simplifies the theoretical
treatment by the possibility to apply the well-known "single-hole problem" to describe the
relevant excitations of the magnetically ordered pseudospin ground state [108, 109]. As
explained in Refs. [106, 110] the strong on-site correlations render very differently for
the case of one electron or hole moving in the AFM background. The polaronic model
constructed by K. Paerschke [106, 110] was used to calculate the dI/dU spectrum [106].
Fig. 4.4.1 (a) display how the charge excitation produced on positive(+) and negative(-)
sides of the dI/dU spectra has different degrees of freedom. The Hamiltonian that describes
its motion is:

$$\mathcal{H}^{+,-} = \mathcal{H}_{\mathrm{mag}} + \mathcal{H}_t^{+,-}, \tag{4.4.1}$$

Here the low energy excitations of the AF $J_{\mathrm{eff}} = 1/2$ ground state are included in $\mathcal{H}_{\mathrm{mag}}$.

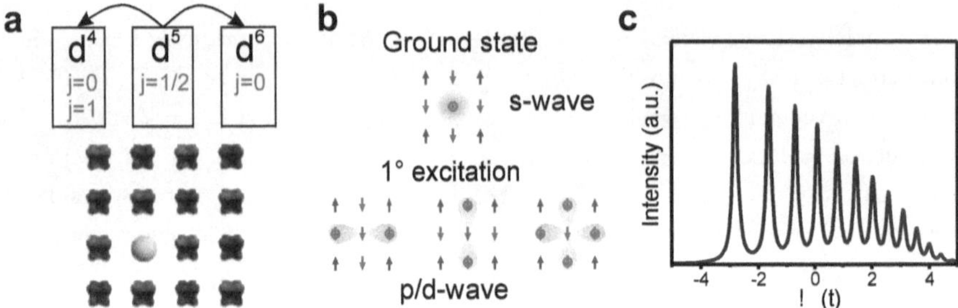

Figure 4.4.1.: (a) depicts a charge in the $j_{eff} = 1/2$ antiferromagnetic background of Sr_2IrO_4. The panel shows how the structure of the spin-polaron is different if a charge is added or subtracted from the background. When an electron/hole is added the spin-polaron has a d^4 electronic configuration with singlet $J = 0$ and triplet $J = 1$ character. In the case of an added hole, the spin-polaron has singlet $J = 0$ d^6 electronic configuration. The hole in interaction with the antiferromagnetic background possesses well-defined excited states with different orbital character, the wavefuctions of which (Reiter functions) resemble atomic states. These are characterized by the size (in terms of the number of virtual hopping processes) of the spin polaron, the wavefunctions' symmetry, and a discrete *ladder* excitation spectrum. (b) depicts the s-wave-like ground state of the spin polaron and the first excited states corresponding to p- and d-wave-like states. (c) Ladder spectrum of the t-J_z model [17] where the exchange interaction $J_z = 0.2t$ lies in the parameter region relevant for Sr_2IrO_4. Figure adapted from Refs. [106] and [107].

The hopping part of the Hamiltonian, $\mathcal{H}_t^{+,-}$, describes the kinetic energy of the charge coupled to the magnons, which gives rise to the spin-polaron quasiparticle. This low-energy effective model has the same operator structure as the effective polaron model of the t-J model described in section 2.2.2. Interestingly, the more components of our model due to the multiplet structure of the polaron allows for more hopping channels, including free (i. e. not coupled to magnons) hopping between first neighbours. The Hamiltonian and its parametrization were shown to give an excellent quantitative description of the measured ARPES spectra [106].

The proportionality between the tunnelling differential conductance dI/dU and the density of states is exploited to relate this model to the STS results. The dI/dU curve was calculated by S. Sykora and K. Paerschke using the relation:

$$\frac{dI}{dU}(\omega) \sim -\frac{1}{2\pi}\sum_{\mathbf{k}} \mathrm{Im} G(\mathbf{k},\omega), \tag{4.4.2}$$

where the time evolution in $G(\mathbf{k},\omega)$ is determined by the Hamiltonian (4.4.1). Next section shows the comparison between the tunnelling differential conductance calculated from eq. 4.4.2 and the experimental one showed in Fig. 4.3.4.

4.4.2. Spin-polaron in Sr$_2$IrO$_4$

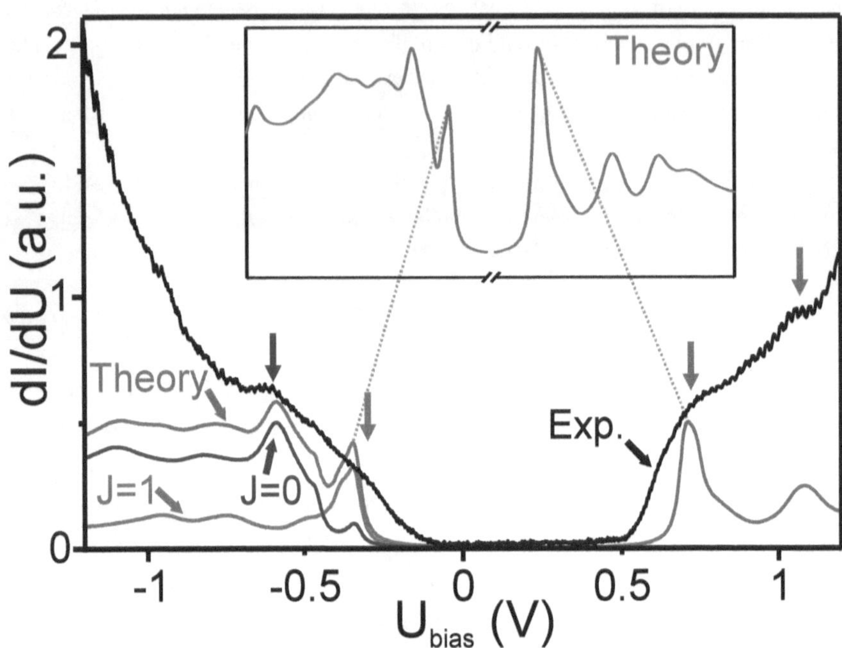

Figure 4.4.2.: Comparison of the theoretical and the experimental differential conductance. The solid green line shows the theoretical spectrum calculated from Eq. (4.4.2) using the momentum-integrated Green's function of the hole/electron. The black line reproduces the experimental spectrum shown in Fig. 4.3.3 (b). For negative bias, the total spectrum is composed of separate contributions due to the multiplet structure of the polaron, i.e. the singlet $J = 0$ (blue line) and the triplet $J = 1$ (red line) contributions. Since the self-consistent Born approximation (SCBA) inherently provides only the distance between excited levels, the calculated results at positive and negative bias are plotted as to match the respective lowest energy anomaly in the experimental data. Furthermore, the experimental spectrum has been scaled along the dI/dU axis in order to achieve a rough match with the calculated results. On the negative side, equidistant features are not present due to dispersive internal degrees of freedom of the charge excitation. Nevertheless, clear spectroscopic features remain discernible. Inset: Calculated total spectrum from SCBA in a wider energy range. Figure adapted from Ref. [107]

Fig. 4.4.2 depicts both the experimental (in black) and the theoretical (in green) differential conductance. The inset of Fig. 4.4.2 display the total calculated spectrum. On the positive side, the ladder-like spectral features are visible. It is highly similar to those of the much more simplistic spin polaron in terms of the t-J_z model (see Fig. 4.4.1(c)). Remarkably this ladder structure present on the positive bias side can straightforwardly be identified in the experimental data by the shoulder-like anomalies at 730 ± 100 meV and 970 ± 100 meV. The resulting matching between experiment and theory is very precise. Especially when noting that the tunnelling spectra are subject of significant broadening, presumably by electron-phonon scattering and higher-order tunnelling processes. Furthermore, the calculation does not include free parameters.

The differences between electrons and holes moving in the $J_{eff} = 1/2$ AFM background (4.4.1) is evident when looking at the negative bias side of the differential conductance. At this site, the ladder spectrum is not present; this is because the motion of the spin-polaron on the negative voltage is very complicated. As previously discussed, the charge excitation has internal degrees of freedom which creates additional interacting channels but also provides a possibility for a nearest-neighbour free hopping of the spin-polaron. For this reason, the spin-polaron becomes greatly dispersive. Additionally, there is a substantial transfer of spectral weight to the incoherent part of the one-particle spectrum. These effects explain that after the momentum summation the dI/dU on the negative bias side (black line in Fig. 4.4.2) is more complicated than the ladder spectrum. However, it is still possible to study the different contributions to the total Green's function which are carried by spin polarons of the two different values of the full quantum momentum, $J = 0$ and $J = 1$. From the calculated spectrum, it is possible to resolve how the singlet and triplet polarons contribute separately to the most salient spectral features. Fig. 4.4.2 depicts the calculated contributions in blue and red for $J = 0$ and $J = 1$, respectively. Therefore, it can be seen that the lowest energy sharp peak has singlet character, whereas the peak at higher energies arises from the triplet polaron. With this information the observed experimental features at about -300 meV and -600 meV are assigned to a primary singlet and primary triplet character, respectively.

Here, the features observed in the experimental spectrum to features of the spin-polaron model have been successfully explained, including the differences in the positive and negative side.

Fitting the spectra

In this section, a statistical analysis is used to find the position of the spin-polaron ladder spectra energy levels. To determine the position of the peaks corresponding to the spin polaron a fitting function $F(x) = F_B(x) + \sum_i F_i(x)$ is applied to the measured spectrum. This function consists of a bosonic background:

$$F_B(x) = \frac{a}{(e^{b/x} - 1)} \tag{4.4.3}$$

and, Gauss functions for the intensity of the peaks,

$$F_i(x) = A_i \exp\left(\frac{-(x - \bar{x}_i)^2}{2\sigma_i^2}\right), \tag{4.4.4}$$

where A_i, $\bar{x}_i$, and σ_i are independent parameters for each peak. The position of the spin-polaron peaks is given by $\bar{x}_i$.

Function (4.4.3) is a phenomenological description of a bosonic background, which is introduced to simulate the dissipation of the energy transferred to the system via tunnelling current. It is assumed that the background is described by a system of bosons (the elementary excitations are magnons). In such a way, the background contribution to the tunnelling response appears like a bosonic distribution function, where the temperature plays the role of dissipated energy which is set proportional to the bias voltage. In addition to this background, the intrinsic excitations of the spin polaron are fitted by Gauss functions where the corresponding positions, amplitudes, and widths are extracted.

The fitting procedure for different spectra is shown in Fig. 4.4.3.

For the spectrum (a) in Fig. 4.4.3 three Gaussian peaks are found, with the fitting parameters given by the following table:

Spectrum (a)		
Background	a	277.594 (a.u.)
	b	6.416 (meV)
Gaussian peak 1	Mean	0.730 (meV)
	σ	0.100 (meV)
	Amplitude	0.500 (a.u.)
Gaussian peak 2	Mean	0.966 (meV)
	σ	0.100 (meV)
	Amplitude	0.500 (a.u.)
Gaussian peak 3	Mean	1.403 (meV)
	σ	0.031 (meV)
	Amplitude	0.828 (a.u.)

The spectra (b), (c) and (d) in Fig. 4.4.3 were taken under the same tunnelling conditions. They do no present the higher energy peak seen in the spectrum (a) at 1.4 eV.

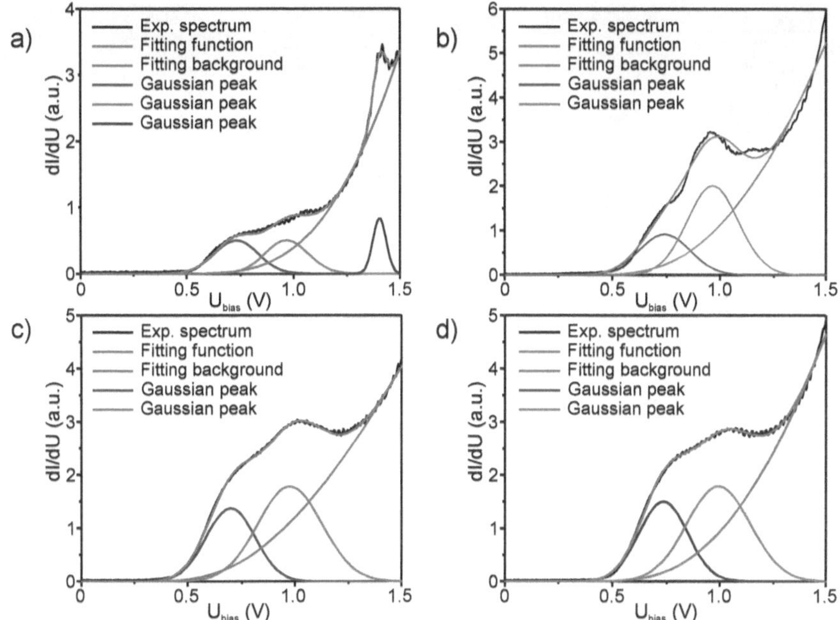

Figure 4.4.3.: Fitting procedure to find the position of the spin-polaron peaks of the spectra (a) from Fig. 4.3.4 and (b),(c) and (d) from a different area.

Since this peak is absent, only two Gaussian peaks are used in the fitting procedure (for the spin-polaron ground state and first excitation). With fitting parameters:

Spectrum (b)		
Background	a	100.0 (a.u.)
	b	4.502 (meV)
Gaussian peak 1	Mean	0.740 (meV)
	σ	0.118 (meV)
	Amplitude	0.910 (a.u.)
Gaussian peak 2	Mean	0.965 (meV)
	σ	0.120 (meV)
	Amplitude	2.000 (a.u.)

Spectrum (c)		
Background	a	35.058 (a.u.)
	b	3.4 (meV)
Gaussian peak 1	Mean	0.698 (meV)
	σ	0.113 (meV)
	Amplitude	1.373 (a.u.)
Gaussian peak 2	Mean	0.975 (meV)
	σ	0.150 (meV)
	Amplitude	1.780 (a.u.)

Spectrum (d)		
Background	a	100.0 (a.u.)
	b	4.675 (meV)
Gaussian peak 1	Mean	0.735 (meV)
	σ	0.113 (meV)
	Amplitude	1.502 (a.u.)
Gaussian peak 2	Mean	0.992 (meV)
	σ	0.149 (meV)
	Amplitude	1.786 (a.u.)

The background parameter a is the amplitude of the bosonic background and b has the role of the bosonic energy in the system. The similar values of b for the different spectra indicate uniform applicability of our used background fitting function (4.4.3). Further consistency of our fitting method is shown by very similar positions of the same peak in different spectra. Many-body phenomena could cause a slight variation of the peak width within the same spectrum.

Electronic correlations in Sr$_2$IrO$_4$

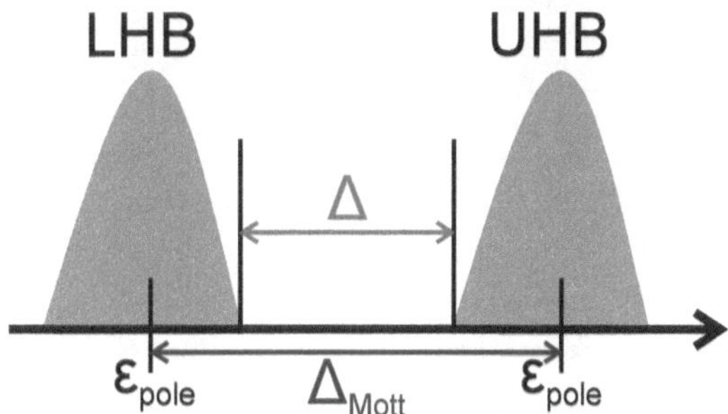

Figure 4.4.4.: Identification of Δ_{Mott} defined as the distance between the LHB and UHB and the effective Δ ($\frac{dI}{dU} \approx 0$) measured in the tunnelling conductance.

The fine features of the tunnelling spectrum of Sr$_2$IrO$_4$ have been solidly assigned to essential spin polaron physics. It is possible to further characterize the electronic correlations in the material by extracting the value of Coulomb repulsion U. Here, U is connected to the Mott gap value Δ^{Mott} as

$$\Delta^{\mathrm{Mott}} = U - \epsilon_{\mathrm{hole}}^{\mathrm{pol}} - \epsilon_{\mathrm{electron}}^{\mathrm{pol}}, \tag{4.4.5}$$

where $\epsilon_{\mathrm{hole}}^{\mathrm{pol}}$ ($\epsilon_{\mathrm{electron}}^{\mathrm{pol}}$) is the binding energy of the polaron formed when a charge (hole or electron) is added to the ground state of the system.

The polaron binding energies are estimated from the SCBA calculations (calculation by K. Paerschke) by setting the hopping part of Hamiltonian Eq. (4.4.1) to zero separately for positive and negative bias cases. In this way, the polaron is artificially fully localized, and its spectral function is simply a delta function. The binding energies are then given by a relative shift between these delta function peaks and the quasiparticle peaks of the full calculation (Fig. 4.4.2). From such consideration, the particular energy values are estimated to be

$$\epsilon_{\mathrm{hole}}^{\mathrm{pol}} = 0.57\,\mathrm{eV}, \quad \epsilon_{\mathrm{electron}}^{\mathrm{pol}} = 0.81\,\mathrm{eV}, \tag{4.4.6}$$

Then the Coulomb repulsion U takes a value between 2.05 eV and 2.18 eV.

4.4.3. Discussion

In this section, the important role of the $J_{eff} = 1/2$ AFM background in the physics of Sr_2IrO_4 has been experimentally confirmed. The existence of spin-polarons has been observed and its internal structure measured. This work supports the similitude with the cuprates.

The data observed for the positive bias voltage has been attributed to the electron doping of the AF background. The present data reveal clear-cut signatures for the prototypical spin polaron ladder spectrum, i.e. the fingerprints of the internal degrees of freedom of the electron being confined within the AF background. Consequently, this study has revealed another similarity between Sr_2IrO_4 and La_2CuO_4 concerning electronic correlations, which hints the possibility that the electron-doped regime of Sr_2IrO_4 at higher doping levels – in analogy to the hole-doped cuprates [73] – could exhibit similar phenomena as the hole doped cuprates. Indeed, the reported signatures of a d-wave gap [78, 79] and stripe-like correlations [80] indicate a phenomenology that may be back traceable to the spin polaron physics [92, 111]. On that account, this study strongly supports the experimental and theoretical efforts [64, 73, 76, 78–80] to find the route to unconventional superconductivity in this material.

The situation in the hole-doped regime is, however, less clean, due to the more complicated physics in the regime. The situation is not analogous to the cuprates because the spectral features are dominated by both singlet and triplet polarons, with singlet states being of lower energy with respect to the Fermi level. Nevertheless, it is not less interesting and an intricate and fascinating doping evolution governed by the interplay of the singlet and triplet polarons may be expected in the hole-doped regime, if chemically achievable. Our study points out that spin-polaron physics is ubiquitous in all families of Mott insulators, suggesting the necessity of studies of high energy spectra of other iridates [112] and cuprates systems [113]. Lastly, it should be pointed out that the observations described here have been supported by ultracold-atom studies of the doped Fermi-Hubbard systems. Koepsell and co-authors have observed signatures of the spin-polaron and its internal structure in a doped Fermi-Hubbard system [114].

4.5. Impurity effects on the LDOS

The resonance peaks of the excitation bands of the ground states have been identified from the tunnelling spectra taken on the clean area. Until here, the effects of defects on the LDOS have been ruled out. However, these effects carry important information for the metal-insulator transition [93, 94] in antiferromagnetic insulators. Therefore the impact

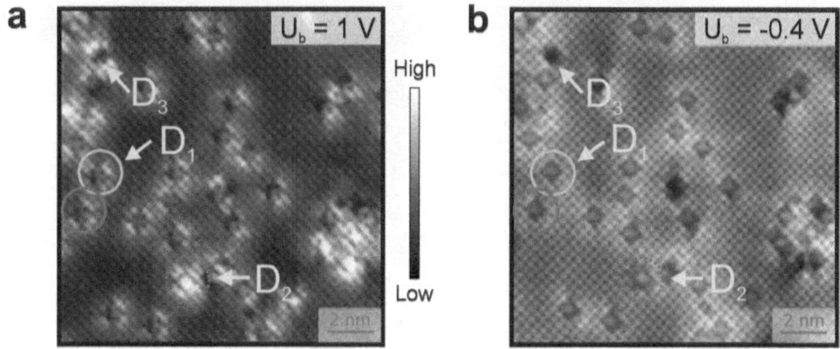

Figure 4.5.1.: (a and b) 14 nm × 14 nm topography of the sample surface measured
at bias voltage wit tunnelling current $I_T = 200$ pA and $U_{\text{bias}} = 1.0$ V(a)
and $U_{\text{bias}} = -0.4$ V (b).

of the primary defect types on the tunnelling conductance is addressed.

4.5.1. Experimental data

Topographic data

Several types of crystalline defects are identified on the sample surface. The three most
abundant types are labelled in Fig. 4.5.1. The type-1 (D_1) and type-2 (D_2) defects are
located on the apex of the oxygen or the Ir position. Type-3 (D_3) defects are located on
the Sr lattice on the surface.

D_1 and D_2 defects show different topographic features for distinct tunnelling conditions
(Fig. 4.5.1). When the defects are present only for the second nearest Sr neighbours,
the individual atoms are distinguishable. The first nearest neighbours are affected by
the presence of the defect and cannot be recognized. However, it is possible to localize
the position of the defects, they lay in between four Sr atoms, in the centre of the IrO_6
octahedra (Fig. 4.5.2 (d)). Furthermore, as shown in Fig. 4.5.2, a careful examination
in the LDOS of the second nearest neighbours around the defects shows chiral behaviour.

Two types of chiral behaviour are enhanced in Fig. 4.5.2 (a and b) to better appreciate
their chirality. The position of the defect is marked in green, and the Sr atoms are shown
with red points. Fig. 4.5.2 (c) display the height profile of the second nearest neighbours.

In both cases, the circumference is navigated anti-clockwise around the defects. The
height profile of the Sr atoms alternate from high to low with inverted pattern if the pro-

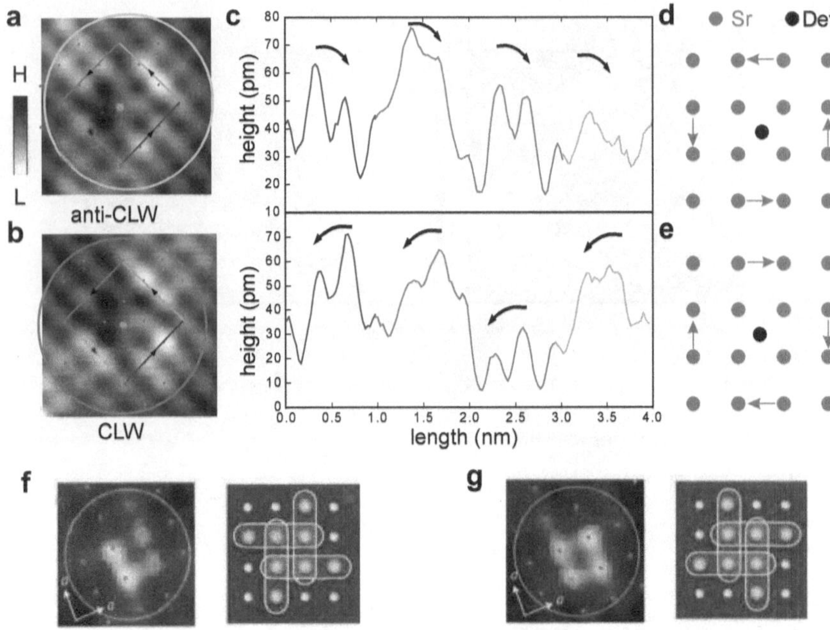

Figure 4.5.2.: Chirality behaviour of the defect topographies and their bias dependencies. The defect topographies are taken from Fig. 1 (b). (a) and (b) depicts the anti-clockwise and clockwise behaviour around the D_1 type defects, respectively. Where the red dots are the positions of the surface Sr atom. In (b), the line profiles of the topography marked in a and b are plotted. The arrows serve as guide to highlight the change of the profile height. (d and e) show the illustrations of the D_1 chirality shown at the defects position in the topography image in panel (a and b), respectively. The top layer Sr atoms are plotted in brown and the central defect in black. (f and g) Show square-shaped defects in $Sr_3Ir_2O_7$ with two different chiralities extracted from Ref. [115]. A schematic representation is displayed to better illustrate both chiralities. The topographic lattice site are indicated by dots.

file is recorded clockwise or anticlockwise. Therefore the corresponding second nearest neighbour pattern is labelled anti-CLW or CLW. The chirality is explained with the rotation of the IrO_6 octahedra with the centre in the defect position. The similar compound, Sr_3IrO_7 has a double-perovskite crystal structure and in-plane rotation of the octahedra showed a similar chiral feature in STM measurements [115].

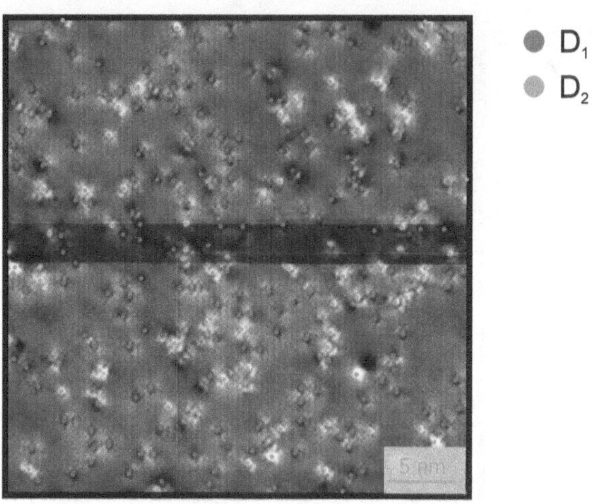

Figure 4.5.3.: 35 nm × 35nm topography measured with $I_T = 200\,nA$ and $U_{Bias} = 1\,V$. The points mark the position of D_1 and D_2 defects in green and red, respectively.

D_1 and D_2 are situated in the Ir/apical oxygen sites. In Fig. 4.5.3 the content of these defects with respect to the Ir atoms is 1.6% for D_1 and 0.7% for D_2. Interestingly, the observed impurity amount in the topographic data shown in Fig. 4.5.3 is consistent with an oxygen deficiency δ of about 2% [72], while it is reasonable to attribute defects D_1 to missing apical oxygen. Other possibilities, such as the Ir atom is replaced by a Sr atom cannot be excluded, and further studies are needed to confirm their chemical origins.

4.5.2. Spectroscopy on defects

Fig. 4.5.4 depicts the changes on the tunnelling conductance for defects D_1 (blue), D_2 (green) and D_3 (purple) or comparison the spectrum taken in the clean area is plotted in black. The first straightforward observation is the reduction of the Mott gap Δ_{Mott} to a value of less than 150 meV for D_1 and D_2. On the other hand, the D_3 shows no evident change to the spectrum taken on a clean place. More precisely, for the D_1 defect the spectral weight below Fermi level is shifted up compared with the one taken at the clean place. Also, there are peak-like features developed at about -200 meV and 250 meV. Although there is spectral weight inside the gap, the spectrum is still gapped with Δ about 100 meV. For D_2, the spectral weight goes inside the gap continuously without developing

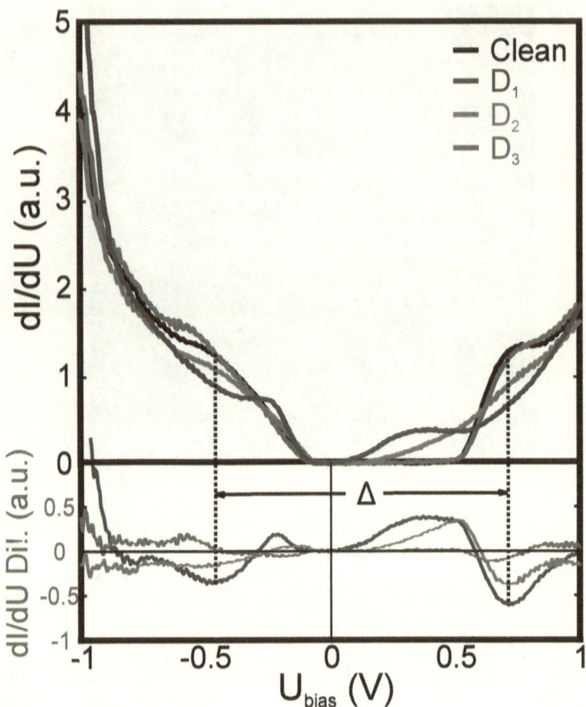

Figure 4.5.4.: Conductance tunnelling spectra taken on different types of individual defect and an area free of defects. The conductance measured on the defects D_1 (blue), D_2 (green) and D_3 (purple) defects compared with the clean area spectrum (black). Lower panel: The spectra differences between the one taken on the clean area and the D_1 (blue), D_2 (green) and D_3 (purple) defects spectra.

features. Δ is slightly higher than for D_1, about 150 meV.

SI-STS observe the spatial extension of D_1 and D_2 defects. Fig. 4.5.5 depicts the evolution of the spectra along two paths on the sample surface (the paths are indicated in rad and white lines in 4.5.5 (a)). Fig. 4.5.5 (c) plots the spectral changes along the white line. The line crosses two individual defects, and the spectra show that the defect effect is very localized and that the Mott gap is not fully closed. This is better seen in 4.5.5 (c), where the spectra line profile is plotted using a colour scale. The extension of the effects of the defect has estimated areas of 2 nm and 1.5 nm diameters for D_1 and D_2

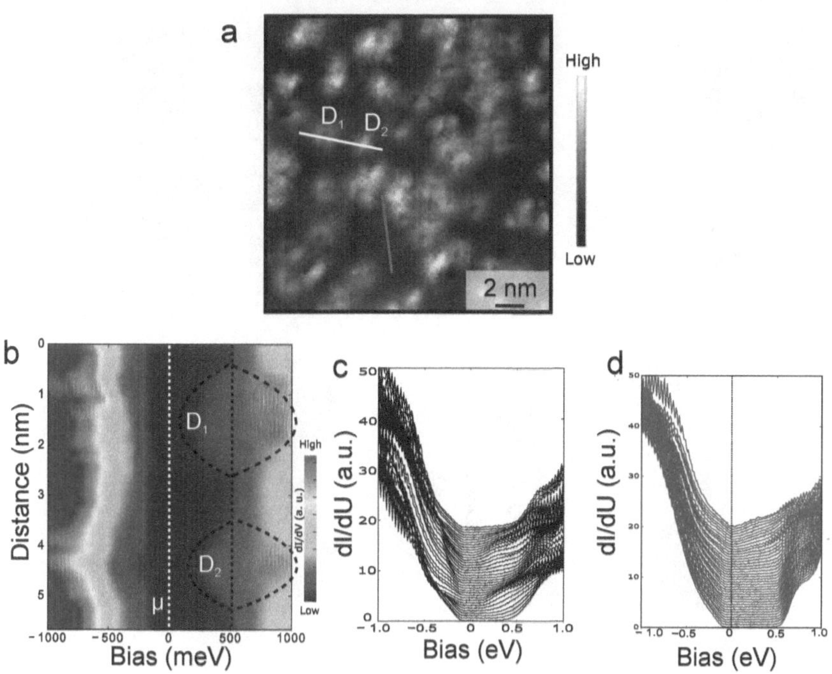

Figure 4.5.5.: The spatial electronic effects of different defects. (a) The corresponding topography of the spectroscopic mapping. Set-point: $V_b = 1.0$ V, $I = 200$ pA. (b) The spectra line profile of the white line show in a with a constant offset for clarity. (c) The conductance spectra profile taken on two different types of defect. (d) the conductance spectra along the dark-red line profile.

defects, respectively. In both cases, there are no states at the Fermi level, indicating that the ground state is insulating. 4.5.5 (d) (red line) tracks the spectra from a clean area towards a cluster with large defects density. Here the spectral behaviour has dramatically changed. In the cluster with high defect concentration, the gap is almost closed, which indicates that in this area charge transport can happen, suggesting a possible percolative charge transport.

4.5.3. Percolative charge transport

Percolation is discussed in the context of charge transport when disorder is introduced into an insulating material and separated metallic and insulating phases appear at the nanoscale

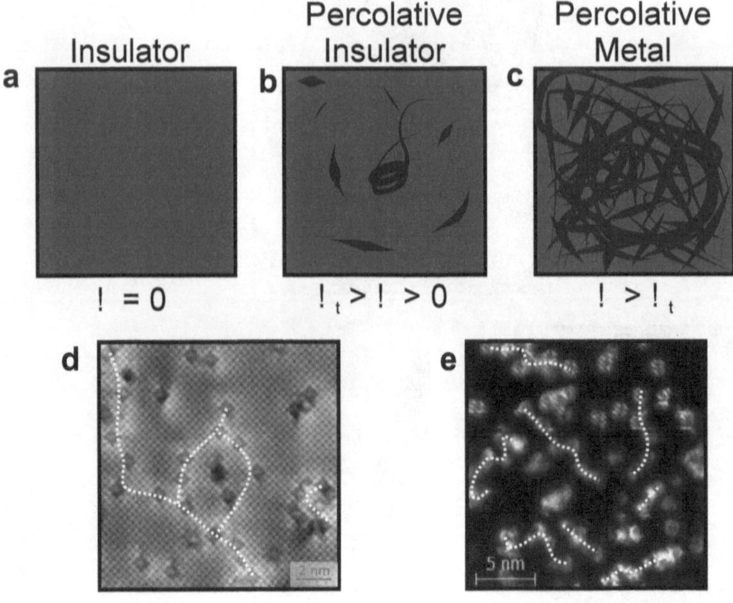

Figure 4.5.6.: (a-c) MIT as a function of δ (defect concentration), with δ_l the percolation threshold. Illustration of a possible percolative path of charge transport in topography (d) and conductance map (e) measurements (indicated by a white dashed line).

region [116, 117]. The percolation threshold is the likelihood that any given region of the medium is sufficiently well connected to the rest to be available for conduction [118]. This mechanism has been proposed as a candidate for the Mott insulator into the non-Fermi liquid transition at half-filling [119, 120].

It has been shown that introducing dilute oxygen vacancies into $Sr_2IrO_{4-\delta}$ turn it into a metallic phase [72]. Based on variable range hopping analysis of resistivity data, the authors speculated with a percolative origin of the charge transport. This observation can be connected with the reduction of Δ in the apical defects observed by STS here. Furthermore, ST-STS showed that the extension of in-gap states is about 2 nm. Patches of defects can produce conductive patches when the defects conglomerates. In agreement with the spectroscopy data,these effects are observed in topography as enhancement of intensity around the defects. Fig. 4.5.6 follows the paths of higher conductance in topography (integrated density of states) and conductance map inside the gap. The white lines offer the possibility of percolative transport.

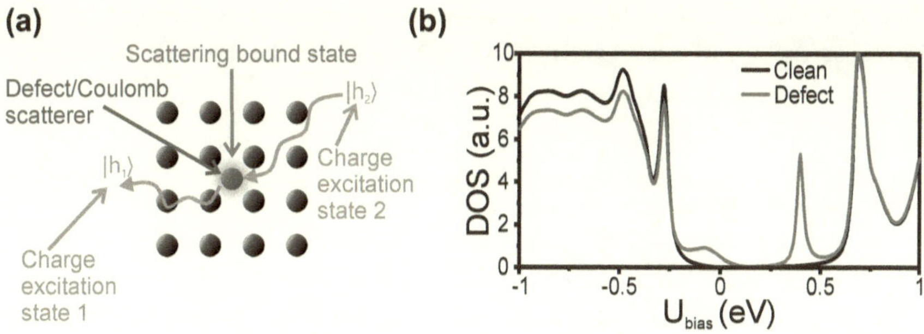

Figure 4.5.7.: (a) Sketch of electronic scattering due to a defect in the lattice and the formation of a bound state (b) Calculated effect of a local defect to the differential conductance. Spectrum for the impurity bound state, in purple, is characterized by an on-site Coulomb scattering potential with a strength of V_{imp}=0.028 eV. The corresponding differential conductance calculated from the clean system is shown in black (Courtesy of S. Sykora). Figure adapted from [121].

Percolation threshold

Following the suggestion of a percolative origin of the charge transport in the STM data S. Sykora [121] performed calculations to find the threshold at which the sample becomes conducting. Using a 2D square lattice with 4 lattice constant as bound size. Which was extracted from the conductance maps give a maximum of 2 nm for the defect bound state. The percolation threshold found in the calculation is 3.7% defect concentration over the Ir atoms. The number matches transition found in the resistivity for samples with of oxygen vacancies [72].

4.5.4. Impurity bound state

Additionally, the effects of defects in the differential conductance were calculated using a standard t-matrix approach (Calculations performed by S. Sykora).

To calculate the Fourier transformed local density of states a local defect is introduced to the one-particle Green's function $G_0(\mathbf{k},E)$ from Ref. [106] acting as a local scatterer with an assumed potential $V_{\mathbf{k},\mathbf{k}'} = V_{imp}$ in momentum space. The result evaluates the momentum-dependent variation of the density of states $\rho(\mathbf{q},E)$ due to the presence of the impurity. This quantity is proportional to the experimentally probed differential conductance dI/dV.

The result for $\rho(\mathbf{R}_i = 0,E)$ as a function of $E = U_{bias}$ is shown in Fig. 4.5.7 (b) (purple line). Here the scattered state tends to localize due to the potential of the defect V_{imp}. For comparison, the calculated spectrum without impurity is shown (black line).

4.5.5. Discussion

The electronic effect of intrinsic defects in Sr_2IrO_4 has been investigated. The main defects on the surface have been located to the apical oxygen or Ir site in the IrO_6 octahedra. The defect estimation (2%) is in good agreement with the measured resistivity for samples with 2% of oxygen vacancies [72], pointing to a vacancy of the apical oxygen as responsible for these defects. STS measurements have observed in-gap states in a confined area around the oxygen defects. In regions where the concentration of defects is high overlapping of the bound states, and partial closing of the gap have been found. Percolation has been proposed as a microscopic mechanism for metallic behaviour. The mechanism is supported with strong evidence of filamentary conductivity through conglomerates of defects. A 2D percolation model has been used to find a limit of 3.7% of defects for the metal to insulator transition. Interestingly, in the study of Sr_2IrO_4 with reduced oxygen content, an exotic metallic state was found [72]. The authors proposed an MIT driven by percolation. The resistivity in the $Sr_2IrO_{4-\delta}$ samples is in agreement with the STM sample (2%). Furthermore, MIT is found at 2% remarkably close to the 3.7% limit estimated here. Additionally, the melting of the Mott state has been studied in La-doped Sr_2IrO_4. The reported pseudogap phase [80] show similar dI/dU than high impurity density regions in Fig. 4.5.5(d). A percolative conduction path may be produced by the La^{3+} impurity potential producing the metal-insulator transition. Spin-polarized STM [122] have found that the inhomogeneous closing of the spectral gap is not related to static short-range antiferromagnetic correlations.

It is interesting to compare with the higher-order Ruddlesden-Popper iridate $Sr_3Ir_2O_7$. A closely related compound with smaller Mott gap of 130 meV measure by STM [115]. A similar closing of the gap has been measured. However, the authors argued that the origin of the gap-filling was related to a defect-induced reduction of the on-site Coulomb repulsion U. They performed band structure calculation and showed that this would cause a rigid shift of the band edges towards each others. In highly doped $Sr_3Ir_2O_7$ percolation have been found to create V-shape pseudogapped regions [120, 123]. The present measurements in Sr_2IrO_4 show a different scenario where the band edges do not move and the spectra are not V-shaped. The reduction of the gap is caused by the in-gap bound states due to defects.

4.6. Summary

This chapter has investigated the motion of charges in the $J_{eff} = 1/2$ AFM background of Sr_2IrO_4, proving the similitude of the novel Mott ground state with the ground state of the cuprates. It has studied the tunnelling conductance in clean areas of the samples and assigned the anomalies in the unoccupied states to the spin-polaron ladder spectrum with excellent quantitative agreement with the self-consistent Born approximation of the polaronic model in the AFM background. Furthermore, the Coulomb repulsion in the material has been quantized, where the U value lays between 2.05 eV and 2.18 eV. Here was presented the first experimental observation of the spin-polarons and its internal excitations, later confirmed by cold-atoms lattice simulators [114]. The observations here support the similitude between the phase diagrams of Sr_2IrO_4 and of the cuprates, such as the observed pseudogap and Fermi arcs at the Fermi level in previous works [64, 73, 76, 78–80].

In the second part, this chapter has also enlightened the doping scheme of the material. Providing atomical resolved nanoscale observations on the defect effects in Sr_2IrO_4. SI-STS has revealed in-gap states (defect bound states) with 2 nm lengths. Additionally, apical oxygen vacancies has been proposed responsible for the more metallic state in concordance with previous studies [72]. Percolation of such vacancies is likely the mechanism behind the MIT in $Sr_2IrO_{4-\delta}$ samples. The percolative model finds a 3.7% threshold in agreement with the experimental critical doping level [72]. Filamentary charge conduction paths have been found in conductance maps and topography. Such scenario is fundamentally different from the usual mechanism of the metal-insulator transition in cuprates. Usually understood as originated from charges moving in the correlated AFM background. This result evidences the differences of the iridates $J_{eff} = 1/2$ with the $S = 1/2$ of cuprates in respect to doping.

5. Iron-based Superconductors

5.1. Introduction

The discovery of superconductivity (at $T_c = 26$ K) in LaFeAsO$_{1-x}$F$_x$ [124] was a milestone in the field of unconventional superconductivity. It marked the beginning of the era of iron-based superconductivity (IBS). Very rapidly, the critical temperature of new iron-based compounds escalated up to 56 K [125] and skyrocketed above 100 K in FeSe monolayer on SrTiO$_3$ [126–128] setting a landmark in the IBS family. The number of similarities between the two families, IBS and cuprates, hinted the possibility of universal trends in high-temperature superconductivity. For compounds with two-dimensional layer structures, where superconductivity emerges from an antiferromagnetic parent compound, competitions between ordered phases are found. However, the microscopic mechanism behind superconductivity remains unknown. One of the key characters in the quest to understand this mechanism is a detailed characterization of the phase diagrams. This study is vital to enlighten the relations between superconductivity, magnetism and electronic orders such as nematicity. The following section intents to give a basic introduction to IBS and more can be found in Refs. [129–134].

5.1.1. Crystal structure

The crystal structure of the IBS parent compounds is tetragonal at high temperatures, with a paramagnetic state. They share a common building-block, a quasi-two-dimensional Fe-As/Se layer, with the As/Se atoms sitting above and below the Fe atoms (see Fig. 5.1.1). The Fe atoms form a square lattice plate. In addition, the As atoms form another square lattice in the Fe-lattice sites, alternating above and below the Fe plane. Therefore, the real unit cell is twice the Fe unit cell, commonly known as the 2-Fe unit cell (5.1.4 (a)).

The members of the IBS are distinguished by the stoichiometric number of their crystal structure. For example the simplest structure FeSe is referred to as member of the 11-family. In more complex structures the Fe-As/Se layers are separated by alkali cations, alkali-earth cations, LnO layer or perovskite-related oxydic slabs. The compounds studied

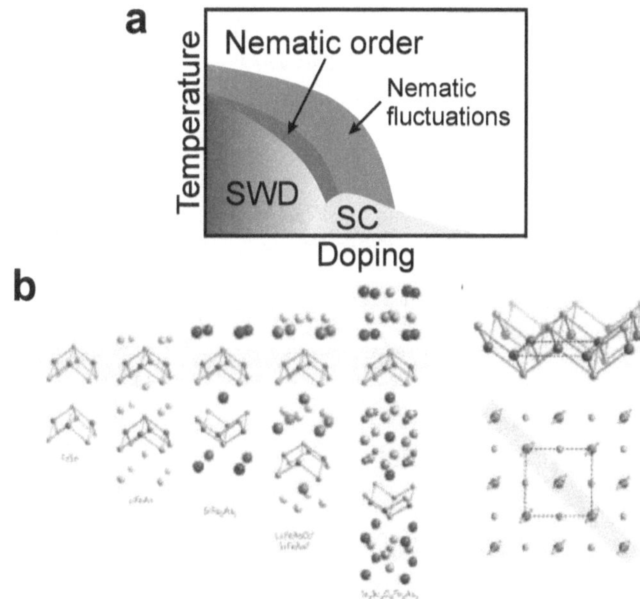

Figure 5.1.1.: General phase diagram and crystal structure of IBS. Schematic phase diagram of iron-based superconductors with nematic phase below T_s (in red) and remaining nematic fluctuations at higher temperatures. Spin density wave (SDW) in the magnetic phase and superconductivity (SC) are marked in blue and yellow, respectively. (b) The crystal structure of five members of the IBS family. The quasi-two-dimensional Fe-As/Se layer is enhanced with Fe atoms in red and As/Se in gold. The arrangement of the spins of the Fe atoms is shown with arrows. Figure adapted from Refs. [34, 130]

in this **book** are members of the 111-family, with main parent compounds; NaFeAs and LiFeAs. In the 111-family the alkali cations of Na/Li sit on top of the As atoms that are positioned below the Fe atoms in the Fe-As/Se layer. For NaFeAs, the Na atoms form a square lattice with lattice parameter $a_{Na} = 2.79\text{Å}$ [135].

5.1.2. Phase diagram

The phase diagram of Co doped NaFeAs (Ref. [136]) is shown in Fig. 5.1.2 as a representative example of the general phase diagram, the compounds of 1111 [137] and 122 [138, 139] families show very similar phase diagrams.

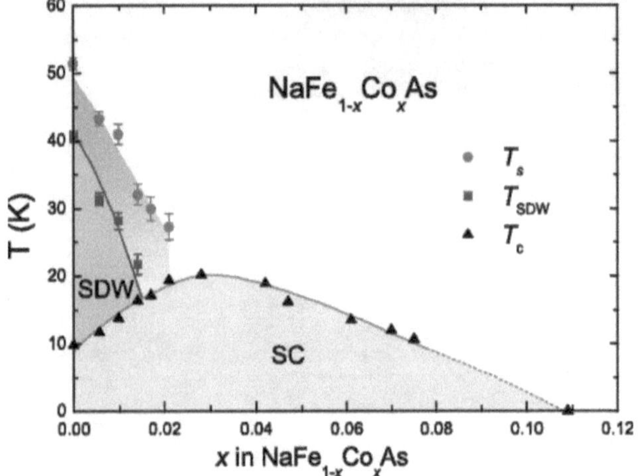

Figure 5.1.2.: Phase diagram Co doped NaFeAs. The nematic phase (in green) below T_s (in red), followed by a magnetic phase (SDW in blue). Superconductivity (yellow) emerges upon Co doping. Figure adapted from Ref. [136]

NaFeAs (the parent compound) undergoes a tetragonal to orthorhombic structural transitions below the transition temperature (T_s). In the orthorhombic phase nematic order sets in. The nematic order is an electronically ordered phase where the x and y direction are non-equivalent (more in section 5.1.5) [32, 140, 141].

Below the structural transition, the systems undergoes a magnetic transition and forms antiferromagnetic order along one of the in-plane Fe-Fe axis (and ferromagnetic order along the orthogonal Fe-Fe axis). The magnetic system is in a SDW phase with AFM vector $Q = (\pi, 0)$ or $(0, \pi)$. Finally the system is turned into the superconducting state by changing the chemical or structural properties, with doping or pressure. In particular, NaFeAs as shown in Fig. 5.1.2 has zero resistivity below 8 K [135], however the superconducting fraction is small (<20%) and it is inferred to be filamentary [142]. Bulk superconductivity is achieved by Co doping, which is substituting Fe directly and introducing extra electrons in the Fe-As/Se layer.

The transition temperatures can be extracted by the analysis of the temperature evolution of the resistivity and its derivative. An example is shown in Fig. 5.1.3 for Co doped NaFeAs. The structural and SDW transition are suppressed upon increasing the Co doping.

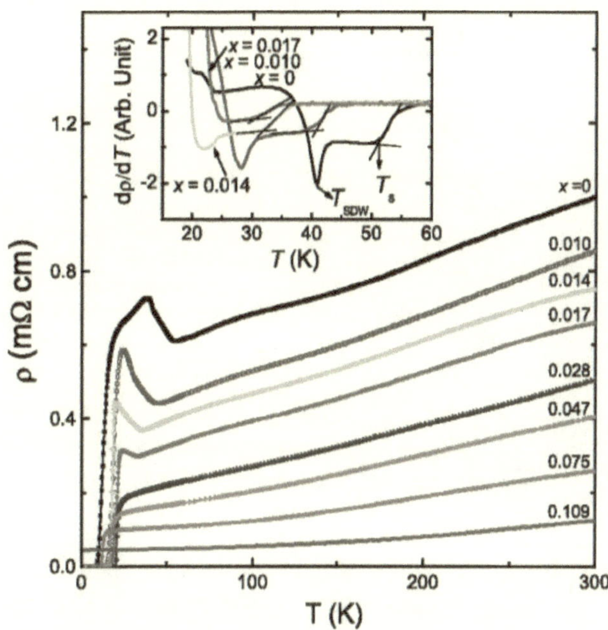

Figure 5.1.3.: Resistivity as a function of temperature for the NaFe$_{1-x}$Co$_x$As single crystals adapted from Ref. [136]

5.1.3. Electronic structure

IBS are multiorbital systems with both electron and hole pockets at different k-points on the Fermi surface (see Figs. 5.1.4 (b) and 5.1.5). For the IBS, the Fe 3d-bands are the only bands close in energy to the Fermi-level and therefore are the only ones accounted in low energy models [143]. In a general picture of the electronic sturcure, the electron pockets are situated at the corner of the Brillouin zone (M point) while the hole pockets

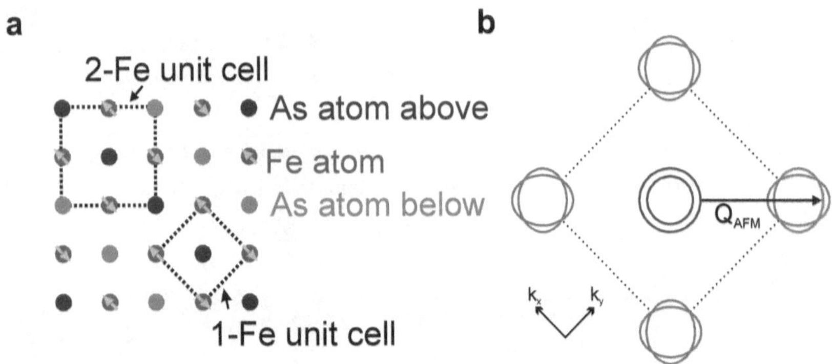

Figure 5.1.4.: (a) 2D structure of the surface Na atoms and underlying Fe atoms. The structural unit cell, one and two Fe atoms per unit cell, are denoted with dashed lines. The yellow arrows indicate the Fe magnetic moment in the stripe AFM order phases. (b) Unfolded Brillouin zone corresponding to one Fe atoms. With a typical Fermi surface consisting of four bands, two hole-like bands at the Γ-point and two electron-like bands in the M points.

are situated at the centre (Γ-point). The number of hole pockets can vary for different compounds. Furthermore, the Fermi surface can be modified by introducing doping atoms [133, 144, 145].

As an example, the electronic structure of NaFeAs (reported in Refs. [146–148]) is shown in Fig. 5.1.5. The low energy band structure of NaFeAs in the tetragonal paramagnetic state (at high temperatures) consists of three nearly circular hole pockets at the Γ point and two elliptical pockets crossing the Fermi level at the M point of the crystallographic two-Fe unit cell (See Fig. 5.1.5). For twinned samples in the nematic phase (45 K), the Fermi surface varies. One of the elliptical electron pockets vanished. However, the one oriented along the antiferromagnetic axis remains present. Additionally, the centre hole pockets become slightly anisotropic. At lower temperatures, the onset of the SDW order has a strong impact on the electronic structure. The SDW doubles the unit cell producing band folding. Hybridization with the backfolded features gaps out large sections of the Fermi surfaces. Thus, the Fermi surface at the Γ point includes four tiny quasi-two-dimensional pockets with Dirac-like dispersions. The pocket along a_{orth} have more significant separation than the ones along b_{orth}. Furthermore, there is a small elliptical pocket at the centre.

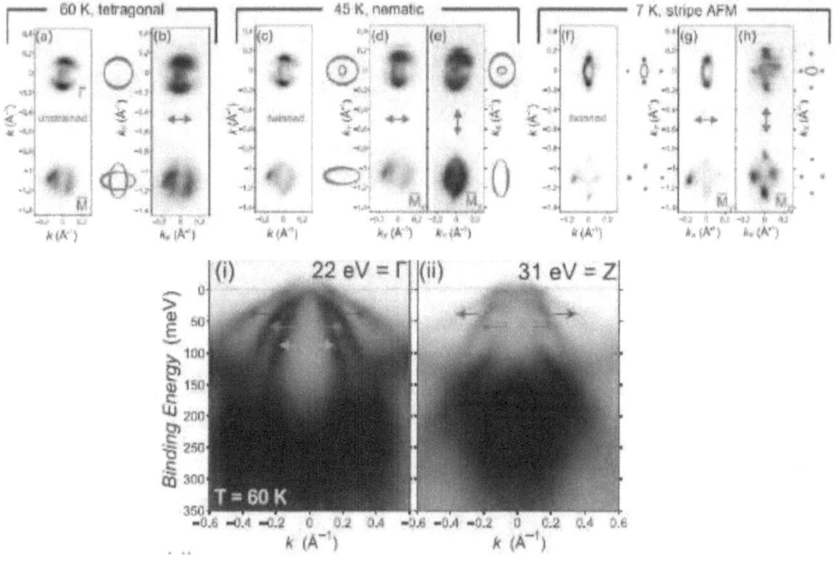

Figure 5.1.5.: Fermi surface maps obtained at 42 eV in LH polarization of (a) unstrained and (b) strained NaFeAs samples at 60 K in the tetragonal phase. Red double-headed arrows indicate the direction of the uniaxial tensile strain on the sample. Schematic Fermi surfaces are drawn in red. Equivalent measurements at 45 K in the nematic phase and (f-h) at 7K in the magnetic phase. (i) High symmetry cuts at 22 eV (Γ) and 31 eV (ii). Data taken from Ref. [146].

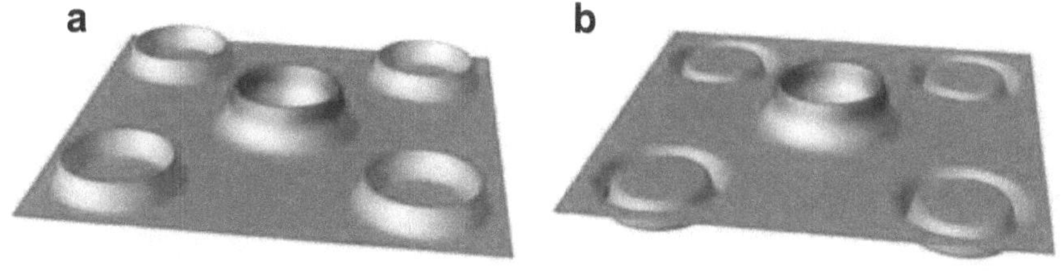

Figure 5.1.6.: Schematic representation of the multi-band s-wave superconducting order parameter for the IBS. (a) two bands s-wave order parameter. (b) $s^{\pm}$-wave order parameter. Figure adapted from Ref. [149].

5.1.4. Magnetism

Upon cooling down the IBS parent compounds, such as $BaFe_2As_2$ or NaFeAs, undergo a magnetic phase transition where an SDW is formed. The magnetic transition has a collinear AFM order with Fe magnetic moment order AFM in on direction and FM in the other (see Fig. 5.1.4 (b)), which can be measured by elastic neutron scattering in $BaFe_2As_2$ [150] or NaFeAs [151]. The ordering vector is $Q_x = (\pi,0)$ or $Q_y = (0,\pi)$ [35, 152]. There are some important exceptions to the general IBS phase diagram: LiFeAs and FeSe are stoichiometric superconductors with no magnetic order. When charges are doped to the system or pressure is applied, the magnetic phase is weakened, as seen in Co doped NaFeAs by neutron scattering [151, 153] or muon spin rotation [154] before disappearing into the superconducting phase. Taking Co doped NaFeAs as example of the magnetic evolution. When Co is doped it replaces Fe and introduces additional electrons to the FeAs layer, and the static magnetic order is suppressed. Additionally, the low energy spin excitations become broader than the spin waves and couple with superconductivity, giving rise to a neutron spin resonance [151], this is also true for Co and Ni doped $BaFe_2As_2$ [155, 156].

The SDW has been attributed to nesting in the Fermi surface. Furthermore, the nested SDW and the metallicity of the parent compounds have been used as pieces of evidences for the itinerant origin of the magnetism in the IBS. In addition, the picture of localized spin bound to the atomic lattice and Heisenberg-type antiferromagnetism has also been proposed. In this picture, some of the orbitals of the Fe orbitals go through a spontaneous symmetry breaking of their occupancy, that triggers magnetic order with the renormalization of the magnetic exchange parameters. Such AFM order can be understood as an effect of the orbital order of the system. A third proposed scenario is the combination of both pictures, localized and itinerant spins, where some orbitals stay itinerant and others present local moment behaviours other [157].

5.1.5. Nematicity

Among the novel quantum phases of matter in the correlated electron systems, the nematic order [30, 158] is of great importance for the Fe-based superconductors and a principal character in this book. By lowering the temperature, the crystal undergoes a structural phase transition where the lattice changes from tetragonal to orthorhombic (T_s). Nematic order is found in the orthorhombic phase (Fig. 5.1.1) [32, 140, 141, 159]. Nematic order breaks the rotational symmetry of the lattice and therefore makes the x and y direction non-equivalent. The order parameter for such transition is called a director (it defines an axis but has no sense of direction) and is borrowed from the study of liquid crystals. In the case of liquid crystals, the nematic state is a consequence of the shape of the molecules or an anisotropic interaction between them. On the other hand, in the electronic nematic phase, the instability is caused by local electronic correlations.

The nematic phase breaks the tetragonal symmetry of the lattice but preserves the translational and spin rotational symmetries. The nematic phase was reported for IBS soon after the discovery of superconductivity [32, 140, 141]. It follows the structural transition and accompanies the AFM phase until superconductivity sets in, involving an electronic competition of the different phases. The detection of the local electronic nematic order is complicated due to a 2-fold degenerate ground state caused by the C_4 symmetry to C_2. A real sample consists of different domains of both ground state, hiding the hints of nematic order. It is possible to break the degeneracy and study the nematic phase to detect such nematic order by applying uniaxial pressure (strain) [159]. A more direct approach is obtained by using local-techniques where STM helped to uncover nematic phases in the cuprates [30] or Fe-based superconductors [32, 33, 160–166]. Furthermore, the nematic fluctuations have also proven to be ubiquitously present over the phase diagram far above the order temperatures of the nematic, AFM or superconducting phases [33, 166, 167].

Like the superconducting transition, the mechanism responsible for the nematic order remains unknown. Since nematic order follows the structural transition, phonons are the first candidate that one can imagine. However, the transition is more likely driven by electronic degrees of freedom. The electronic origin is proven by measurements of the anisotropy of the resistivity using strain [159], where it was shown that the susceptibility diverges near the nematic transition, this probes that the structural distortion is a conjugate field to the primary order parameter (nematic electronic). Various microscopic model for electronic nematicity had been propossed, the two main scenarios put charge/orbital fluctuations or spin fluctuations as drivers of the nematic instability [27, 34] the question remains controversial. As in the case of the magnetic transition, where the orbital-selective spin-fluctuations model have explained the anisotropy in the QPI and dc conductivity [168–170].

5.1.6. Superconductivity

For most of the IBS, superconductivity is achieved as an effect of doping or pressure. Nonetheless, it is possible to find stoichiometric superconductors as a result of distinct Fermi surfaces (LiFeAs [171] and FeSe [172]). In any case, the high superconducting temperatures suggest that the pairing mechanism deviates from the conventional electron-phonon pairing mechanism [173] in favour of pairing from repulsive electron-electron interaction [21, 174, 175].

In order to explain the superconducting instability, two main scenarios have been proposed: spin and orbital fluctuations. The proximity to the magnetic phase supports the spin scenario. Here, the Coulomb interactions between electron and hole pockets can be approximated by a repulsive effective potential ($V_{inter} > 0$) mediated by spin fluctuations [175]. This scenario required a Δ_k that changes sign between the hole and electron pockets with vector Q_{AFM} due to the symmetry of the Fermi surface Δ_k has a $s_\pm$ symmetry (Fig. 5.1.6 (b)). The alternative model that received the most attention is the orbital fluctuations model [174]. Here the system has some orbital order even in the itinerant model. Fluctuations of this order can lead to an attractive interaction between electron and hole pockets ($V_{kk'} < 0$) and therefore form Cooper pairs. Here Δ_k is a conventional state with no sign change inside the pockets therefore s_{++} symmetry (Fig. 5.1.6 (b)).

Orbital or spin fluctuations are likely candidates for the superconducting pairing glue, with anisotropic gap functions and order parameters s^{++} or $s^\pm$, respectively [21, 174, 175]. Moreover, the successful model should explain the origin of magnetism and nematicity in iron-based superconductors [27, 34]. Models with orbital selectivity potentially conciliate both tendencies. Orbital-selective account the differences in the correlation strength for electrons with different orbital character as responsible for the concentration of the pairing in a particular orbital channel [157, 176, 177].

5.2. The 111 family

This chapter is devoted to STM measurements on Li doped NaFeAs, which is a recently discovered member of the $\langle 111 \rangle$ family of IBS. Characteristic of this family is the cleaving plane between adjacent Na/Li layers, which results in charge-neutral surfaces [171]. Consequently, the electronic structure of the surface is equivalent to that of the bulk compound [178]. As a result, this family offers some of the best platforms to study IBS with surface-sensitive methods, e.g. ARPES and STM. STM plays an essential role in the determination of electronic phases, with a combination of information in real and q-space [33, 146, 147, 179, 180].

The unique conundrum of the $\langle 111 \rangle$ family is the different nature of its members when they have Na or Li as the alkali metal atom. LiFeAs is a stoichiometric superconductor. Doping does not enhance superconductivity but suppresses it [171, 181, 182]. A notable difference of LiFeAs with the rest of IBS is the absence of static magnetism [171, 181] or Fermi surface nesting [54, 180]. Furthermore, spins excitations are weak and incommensurate with little changes across the superconducting transition [183]. Nevertheless, nematic fluctuations have been shown responsible for a small-q instability of the system [121]. Moreover, a electronic density fluctuation with small-q has been reported upon uniaxial strain [165].

On the other hand, NaFeAs is a prototypical member of the IBS. It shows an AFM spin density wave phase with Neel temperature T_N=43 K. The onset of SDW order has a strong impact in the Fermi surface, backfolding the electrons and hole pockets into tiny quasi-2D reconstructed pockets [146]. Nematic order develops below the structural transition T_S=54 K [184, 185]. However, nematic fluctuations persist above the transition [33]. The separation between transition temperatures is higher that in other IBS compounds and therefore it makes NaFeAs an ideal candidate to study the nematic phase [33, 146, 186]. Doping suppresses the magnetic order and develops bulk superconductivity [154, 187–189]. The electronic phases are not always antagonist and coexistence is found between superconductivity, nematicity and AFM [189, 190]. Apart from the known phases, pseudogap-like spectral features have been detected in the Co doped compounds [191].

5.3. Li doped NaFeAs

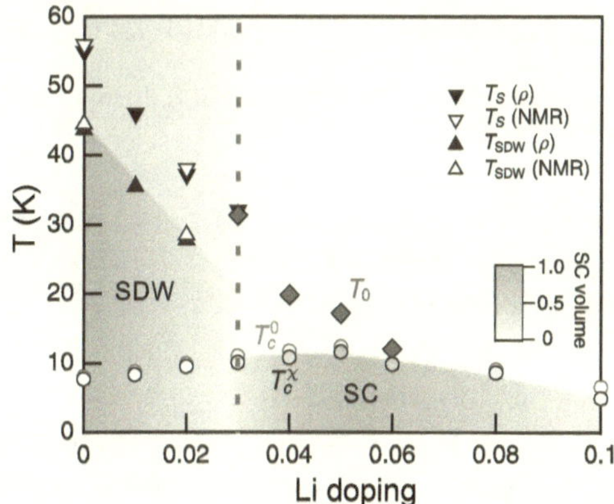

Figure 5.3.1.: Phase diagram of $Na_{1-x}Li_xFeAs$ as a function of temperature and doping. The transition temperatures are measured with transport, magnetization and NMR. The red dashed line denotes a change in the behaviour of the sample with an emerging new phase at T_0. Figure courtesy of S.-H. Baek and adapted from Ref. [192].

The conundrum in the ⟨111⟩ family (differences between NaFeAs and LiFeAs) motivated the investigation of $Na_{1-x}Li_xFeAs$ phase diagram with single crystals. The phase diagram (plotted in 5.3.1) was studied for the first time in Ref. [192].

Fig. 5.3.2 shows data from the single crystal extracted from Ref. [192]. Fig. 5.3.2 (b) displays the variation of the lattice parameter c of the $Na_{1-x}Li_xFeAs$ single crystals extracted from the X-ray diffraction pattern. c decreases systematically with increasing the Li content up to 0.06.

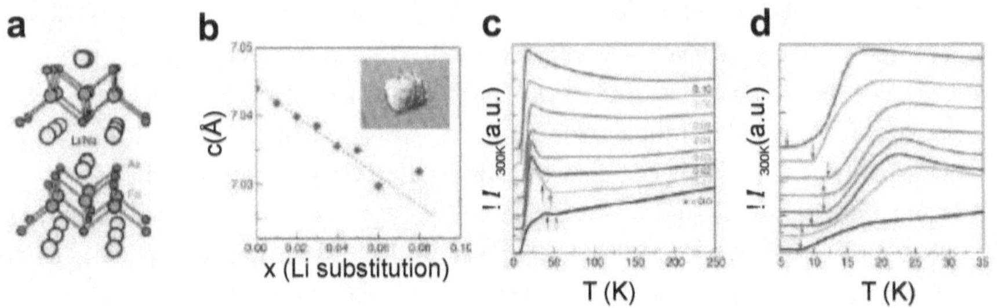

Figure 5.3.2.: Bulk characterization of $Na_{1-x}Li_x$FeAs. (a) $Na_{1-x}Li_x$FeAs crystal structure. (b) characterization of the c-axis for Li doped samples. Inset shows an image of a sample crystal. (c-d) Temperature dependence of the normalized resistivity. in (c) the black and magenta arrows denote T_{SDW} and T_s, respectively. For (d) the evolution of T_c is followed with blue arrows. Figures courtesy of S.-H. Baek and adapted from Ref. [192].

The temperature dependence of the in-plane resistivity was used to determine the transition temperatures T_c, T_{SDW} and T_s (Fig. 5.3.2 (c-d)). As in the case of Co doped NaFeAs, the structural and SDW transition temperatures were determined with the derivative of the resistivity and T_c is defined as the temperature at which resistivity reaches zero because the superconducting transition are round. With increasing the doping, the magnetic and nematic transition temperatures are suppressed for $x \geq 0.03$, interestingly at this doping superconductivity becomes bulk. The highest transition temperature ($T_c = 12.3$ K) is found at $x = 0.05$ doping.

Fig. 5.3.3 displays the NMR data on $Na_{1-x}Li_x$FeAs used to determine the presence of long-range magnetic order (AFM split). Following the behaviour of the parent compound, long-rage magnetic order is confirmed for samples until $Na_{0.98}Li_{0.02}$FeAs. Whereas, for $x \geq 0.04$ there is no signature of SDW order.

For $x \geq 0.03$ the system develops a new a non-magnetic phase transition which is better seen in the spin-relaxation rates. In Fig. 5.3.4 the upturn of $1/T_1T$ is related to the antiferromagnetic instability which persists even for $x \geq 0.03$. Below the transition, the divergence of the spin-fluctuations is suppressed, which is consistent with the disappearance of SDW order. Unexpectedly, however $1/T_1T$ drops fast below T_0 and forms a peak, indicating a phase transition. The temperature is marked as T_0 and appears to follow

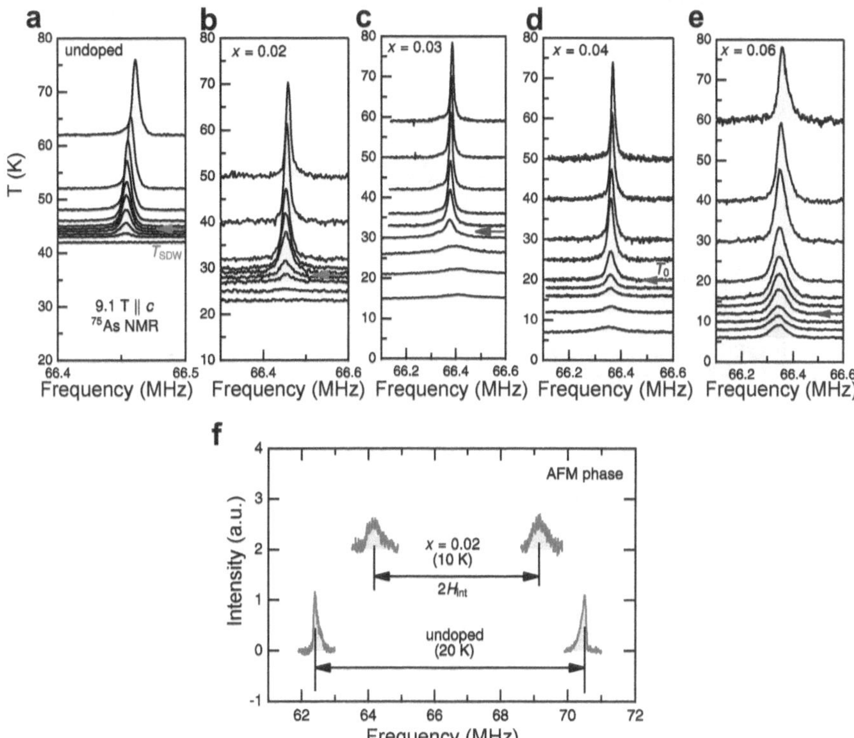

Figure 5.3.3.: ^{75}As NMR spectra for Na$_{1-x}$Li$_x$FeAs. (a-e) Temperature dependence of the central line of ^{75}As NMR spectra for Na$_{1-x}$Li$_x$FeAs for $H\|c$ measured at 9.1 T (displayed for $x = 0$, 0.02, 0.03, 0.04 and 0.06). In (a-b) the central line disappears below T$_{SDW}$ due to the stated antiferromagnetic order in the Fe spin moments. For $x \geq 0.003$(c-d) the line remains finite down to low temperatures, this indicates the absence of long-range magnetic order. (f) AFM split of the ^{75}As lines detected for $x = 0$ and 0.02. Figures courtesy of S.-H. Baek and adapted from Ref. [192].

the structural transition. Spin-fluctuations are enhanced before T_0 and suppressed in the nematic state. At optimal doping, there are no signatures of long-range magnetism. Therefore the phase transition is not spin nematic and was attributed to ordering in the charge/orbital channels. This is a new phase transition, which has not been reported in different dopings. No signs or magnetic order are found for the new phase, nor in the [75]As spectrum or in the spin fluctuations.

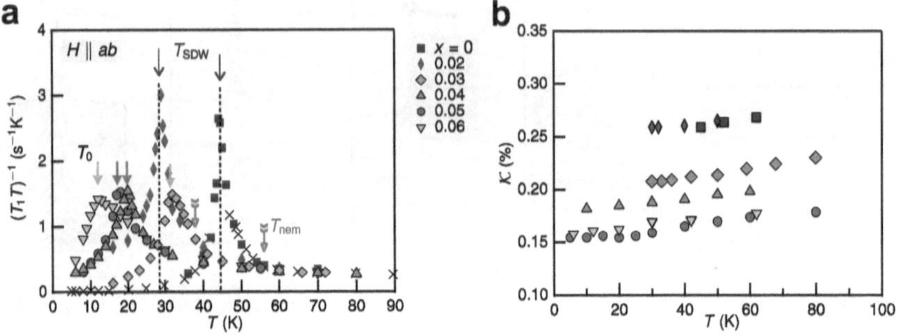

Figure 5.3.4.: Spin-lattice relaxation and Knight shift in $Na_{1-x}Li_xFeAs$. (a) Spin-lattice relaxation rates as a function of temperature and doping. Sharp transitions are identified for the SDW state at $x \leq 0.02$ and T_0 for large doping. (b) The Knight shift as a function of temperature and doping Figures courtesy of S.-H. Baek and adapted from Ref. [192].

All the information is added to the phase diagram, which can be seen in Fig. 5.3.1. Interestingly, the phase diagram can be divided into two regimens, separates by a frontier at $x = 0.03$ of Li doping (the frontier is marked with a red dashed vertical line in Fig. 5.3.1 (a)). In the left side of the frontier, the sample follows the behaviour of the parent compound with a structural transition, a nematic phase and an SDW phase. On the right side, the samples show bulk superconductivity and a different phase transition at T_0. The differences in the phase diagram are supported by the Knight shift, which changes abruptly at $x = 0.03$. The transition corresponding to the spin-fluctuations at T_0 has not been reported in NaFeAs or other dopings [154, 187, 188, 193] or other members of the IBS family. Addtionally, below T_0 the [75]As signal amplitude is notably reduced for both field configurations (see [192]). The suppression of signal intensity indicates that the volume fraction of the sample seen by NMR decreases in the ordered phase. This phenomenon is quite similar to the wipeout effect observed in the charge stripe phase of cuprate superconductors, in which NMR relaxation rates of the nuclei in spin-rich regions become too fast to be detected [194]. The underlying mechanism of the signal wipeout in $Na_{1-x}Li_xFeAs$ remains unclear and needs further investigation. Interestingly, the similar

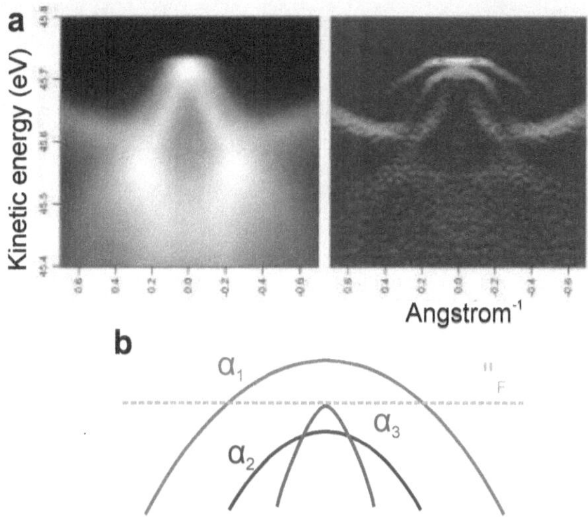

Figure 5.4.1.: ARPES measurements on Na$_{1-x}$Li$_x$FeAs. (a) High symmetry cut at the centre of the Brillouin zone for Na$_{0.96}$Li$_{0.04}$Fe at low temperature (2 K). The image displays the symmetrized sum of the data measured with vertical and horizontal polarization. (b) Schematic band structures along the M-Γ-M path. ARPES data courtesy of Y. Kushnirenko, S. Federov and S. V. Borisenko.

wipeout of the NMR signal was also observed in the ^{77}Se NMR study of FeSe in the nematic phase which does not involve any static magnetism [195].

All this motivated the local study of the new charge/orbital phase order at T_0 in real and momentum space by SI-STM. The work presented in this **book** was performed using high-quality single crystals samples of Na$_{1-x}$Li$_x$FeAs synthesized by Prof. Kee Hoon Kim and collaborators in the Institute of Applied Physics, of Seoul National University. The samples were mounted into the STM inside and Argon box to avoid exposition to oxygen.

5.4. ARPES on Na$_{1-x}$Li$_x$FeAs

Traditionally, the discussion of STM/S has been enriched by the comparison with the electronic structures directly measured by angle-resolved photoemission spectroscopy (ARPES). Here the measured electronic structure of the 0.04 Li doped NaFeAs samples is compared to the well known

Low-temperature ARPES data at the Γ point (along the M-Γ-M direction) of 0.04 Li doped NaFeAs is shown in Fig. 5.4.1 (a). The image shows three hole-like band dispersion, two of them crossing the Fermi level. A sketch of the hole-like band is shown in Fig. 5.4.1 (b) with the bands labelled as α_1, α_2 and α_3. The system does not show the reconstruction related to the SDW in the NaFeAs. The data resembles the one measured in the tetragonal phase of the parent compound.

5.5. Results on $Na_{0.97}Li_{0.03}FeAs$

Here are shown the first STM/S results on $x = 0.03$ $Na_{1-x}Li_xFeAs$. This doping composition is at the frontier between the magnetic and non-magnetic regimes of the phase diagram (Fig. 5.3.1). At this doping, the structural transition coincides with the newly appointed phase transition $T_s = T_0 = 32$ K. The data were obtained at 5 K, below $T_0{=}32$ K and $T_c{=}11$ K.

5.5.1. STM measurements

Topography

Fig. 5.5.1 (a) portray a typical topographic image of $Na_{0.97}Li_{0.03}FeAs$ in an area of 50 nm $\times$ 50 nm measured with 512 px $\times$ 512 px at $T = 5.2$ K. Images of the most common crystalline defects for $Na_{0.97}Li_{0.03}FeAs$ have been enhanced in Fig. 5.5.1 (d-e). The most abundant defects on the surface looks like dark spots. Interestingly, there are two types of dark defects (labelled D_1 and D_3). D_1 produces electronic scattering which enhances the topographic contrast in their surroundings. Those labelled D_3 do not show neighbouring electronic scattering. The second most abundant defect have a dumbbell-like shape; they are labelled as D_2. The lattice orientation can be deduced from the fast Fourier transform (FFT) image of the topography (shown in Fig. 5.5.1 (b)). The Bragg peaks marked in the image along the diagonals correspond to the Na-Na crystal directions. Consequently, D_2 are oriented along the diagonals of the image, and this is a typical feature of the Fe-site defects observed in LiFeAs [54, 121, 179, 196, 197] and NaFeAs [33, 160, 191, 198] (an example is shown in Fig. 5.5.2 (a) where the dumbbell-like defects are marked with yellow lines). In general, these defects are similar to defects reported in previous studies on NaFeAs and LiFeAs, which are shown in Fig. 5.5.2 for comparison. D_1 defects probe enhancing of the tunnelling current in their vicinity, which results in electronic scattering.

A bias voltage dependence study has been performed to investigate the changes in the electronic scattering. The topographic changes for the same field of view can be seen in Fig. 5.5.3, which display three topographic images for an area of 120 nm $\times$ 120 nm with a resolution of 1024 px $\times$ 1024 px. From 50 mV (Fig. 5.5.3 (a)) to -10 mV (Fig. 5.5.3 (b)) there are noticeable changes in the scattering around D_1 or D_2 (examples are marked in orange or black, respectively). The enhanced areas shown in black boxes are shown in Fig. 5.5.3 (d,e), the electronic scattering (in white) changes its locations and overall shape. Furthermore, the image confirms that the position of the topography does not change. There is an abrupt change when the bias voltage is set to -30 mV. At this voltage, the defects are not distinguishable any more.

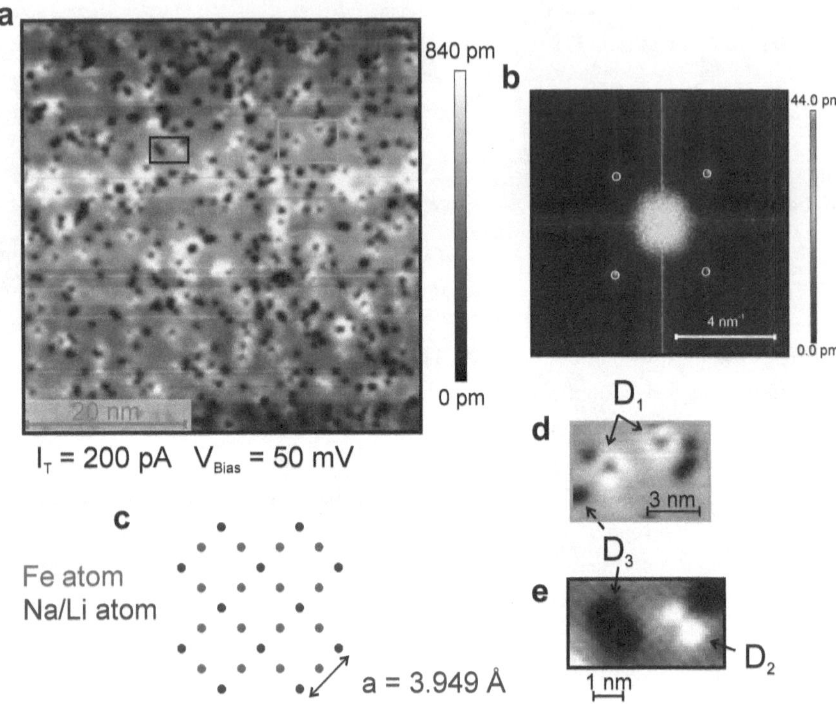

Figure 5.5.1.: Atomically resolved topography of $Na_{0.97}Li_{0.03}FeAs$ measured at $T = 5.2$ K. (a) The topography of an area of 50 nm × 50 nm of the sample surface measured at bias voltage $U_{bias} = 50$ mV and tunnelling current $I_T = 200$ pA. (b) displays the Fourier transform of (a), which clearly shows the Bragg peaks corresponding to the first Brillouin zone. (c) depicts a sketch of the top view of crystalline structure. The lattice vector a is orientated along the diagonals of the topography as extracted from the Bragg peaks. (d and e) show the area inside the orange and black boxes, respectively. The most common crystalline defects of $Na_{0.97}Li_{0.03}FeAs$ are labelled as D_1, D_2 and D_3.

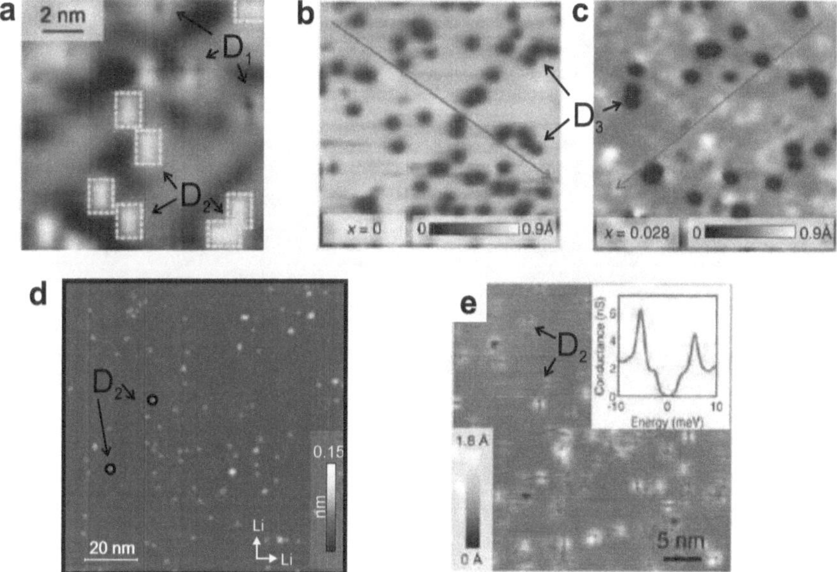

Figure 5.5.2.: STM images of NaFeAs and LiFeAs. (a-b) A constant current image of NaFeAs with atomic resolution. (c) A topography of the optimally doped $NaFe_{1-x}Co_xAs$. Data for (a-c) taken from Ref. [190]. (d-e) Typical topographies of LiFeAs taken from Ref. [166] and Ref. [179], respectively.

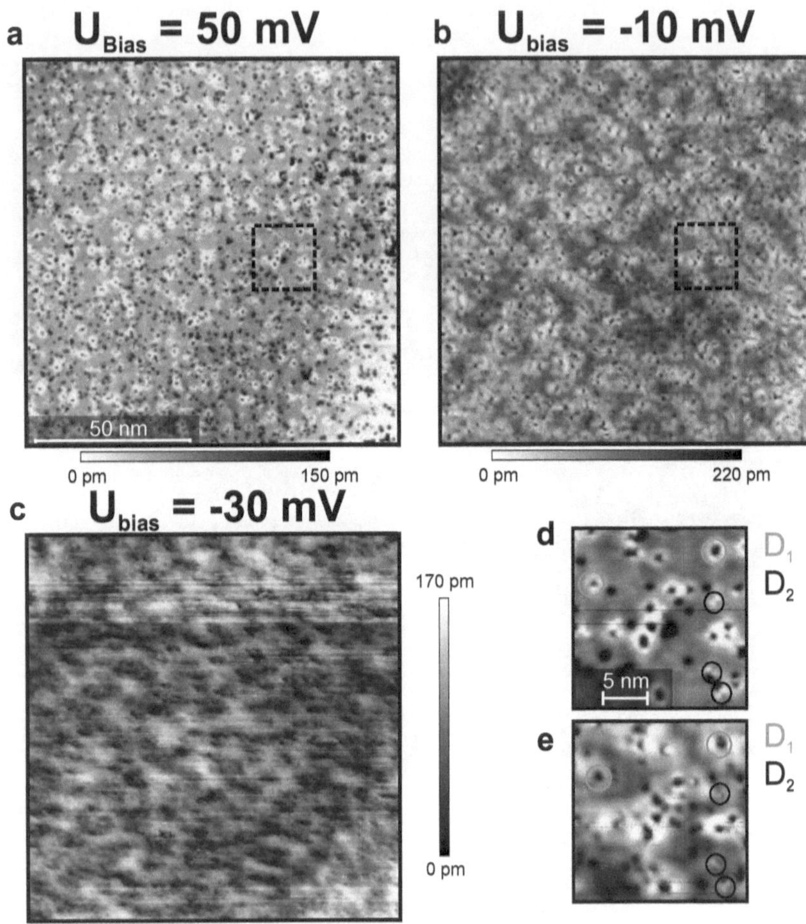

Figure 5.5.3.: Voltage dependence of topographic images of $Na_{0.97}Li_{0.03}FeAs$. (a-c) 120 nm × 120 nm topographic images measured with 1024 px × 1024 px at T=5.2 K with bias voltages 50 mV, -10 mV and -30 mV, respectively. All the images display the same field of view. (d-e) show an enhanced image of the area marked in dash black for (a and b).

dI/dU maps

Small changes in topography could result from changes in the LDOS which should be confirmed by dI/dU maps. Figs. 5.5.4 (50 mV), 5.5.6 (-10 mV) and 5.5.7 (-30 mV) display the corresponding dI/dU to the topographies shown in Fig. 5.5.3 (a-c). The dI/dU images allow to probe changes in the LDOS when the changes in the surface height-profile are negligible.

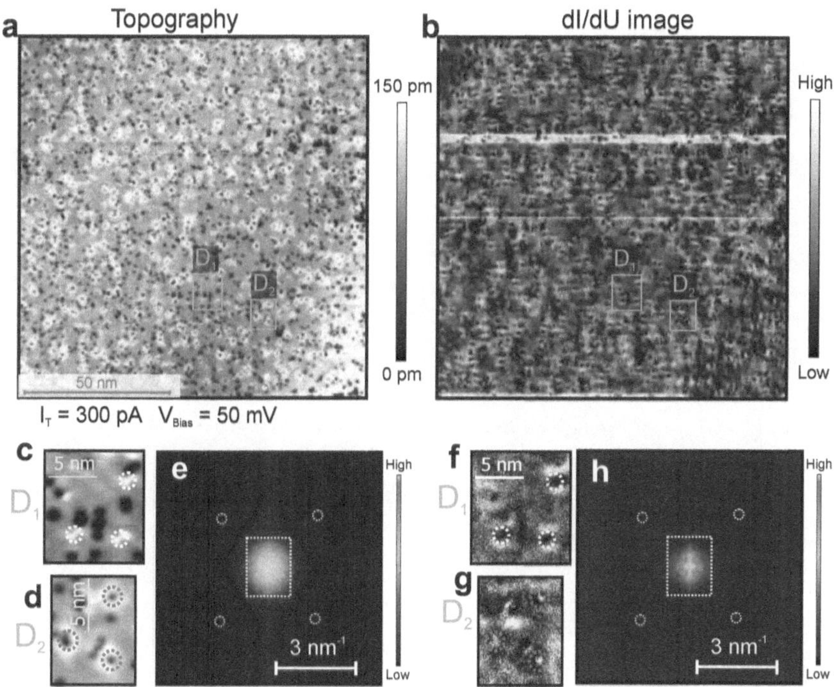

Figure 5.5.4.: 120 nm × 120 nm topographic and dI/dU images of atomically resolved surface measured at $T = 5.2$ K, with bias voltage $U_{bias} = 50$ mV and tunnelling current $I_T = 300$ pA.

Fig. 5.5.4 display the topography (a) and the simultaneously recorded dI/dU image (b) of 120 nm × 120 nm area of the sample surface measured at $U_{bias} = 50$ mV with constant $I_T = 300$ pA. The D_1 and D_2 defects inside the image have been counted to calculate their abundance with respect to Fe atoms. It is 0.008 for D_1 and 0.01 for D_2.

Two areas with D_1 and D_2 defects are enhanced to better appreciate the changes in the dI/dU due to the individual defects. In the topography, Fig. 5.5.4 (c) show three D_1 defects in white and Fig. 5.5.4 (d) depicts three D_2 defects in white. Fig. 5.5.4 (f and g) displays their respective for dI/dU maps. Especially interesting is the dI/dU signal for D_1 (Fig. 5.5.4 (f)), which shows a local breaking of the C_4 symmetry. The structure of the defect has a dark circle in the centre surrounded by two bright horizontal lines. Such an effect is not appreciable for D_2 (Fig. 5.5.4 (g)).

Fig. 5.5.4 (e and h) displays the FFT images of Fig. 5.5.4 (a and b). Following the observation of the nematic D_1 defects, a global breaking of C_4 is corroborated by the FFT of the dI/dU image (Fig. 5.5.4 (h)). The image shows a C_2 symmetric feature in the centre. Contrarily, the topography FFT (Fig. 5.5.4 (e)) does not show a central structure, and only the Bragg peaks are recognizable.

Fig. 5.5.5 investigate the breaking of the C_4 symmetry observed in the dI/dU image from Fig. 5.5.4. Three areas (**1** in Fig. 5.5.5 (b), **2** in Fig. 5.5.5 (c) and **3** in Fig. 5.5.5 (d)) are selected to show the changes in real space produced by D_1 defects, which show local unidirectional features in the sample surface (pointed by two parallel white arrows). The selected areas **1** and **2** present these unidirectional features along the same Fe-Fe direction. Per contra, the features inside the **3** show a 90° rotation with respect to the areas **1** and **2**, which is confirmed by the corresponding FFT images (Fig. 5.5.5 (e-g). **1** and **2** show horizontal elongation in the centre of the FFT image (white arrows are plotted to help the recognition). On the other hand, the central elongation in **3** is horizontal. The boundary between the domains is delineated with a dashed white line in Fig. 5.5.5 (a).

Fig. 5.5.6 shows the same area of the sample surface from Fig. 5.5.4 measured with $U_{\text{bias}} = -10$ mV. As seen before, the topography shows high a abundance of defects.

A difference with Fig. 5.5.4 is that the background appears to segregate into two different types with lighter or darker contrast. An area with a high background (A_1) and two areas with the lower background (A_2 and A_3) are delimited by violet and black rectangles, respectively.

In the dI/dU image (Fig. 5.5.6 (b)) the contrast is inverted with respect to the topography image (Fig. 5.5.6 (a)). The inversion is seen in the general background and the defects features. The dark defects are easily recognized as bright points the dI/dU (five dark defects are chosen in white circles to illustrate the change). Another observation in the dI/dU image is the existence of electronically ordered patches along vertical or horizontal lines for distances of ≥ 5 nm (see conductance images for areas A_2 and A_3). The distance between patches is estimated ~ 8 nm.

Since topography images are proportional to the energy integrated LDOS and dI/dU image only to the LDOS at U_{Bias}. The different contrast for topography and dI/dU image can be understood thinking in a peak of the LDOS at U_{Bias} that is not seen in topography

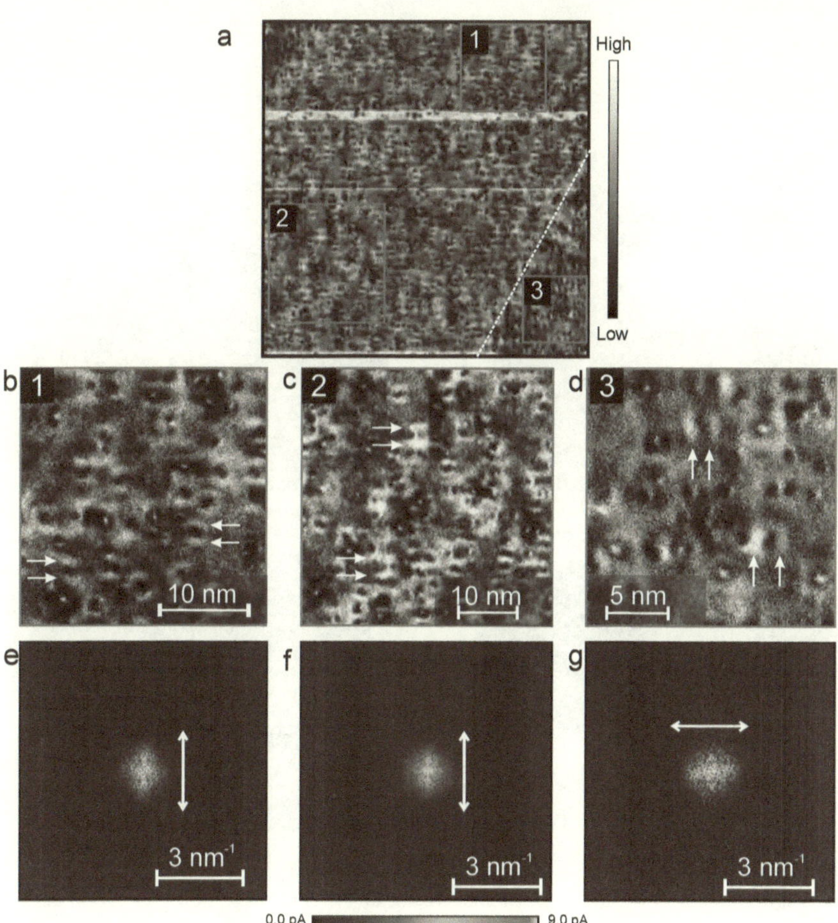

Figure 5.5.5.: Nematic features in dI/dU image. (a) Nematic domains in the dI/dU image taken from Fig. 5.5.4. (b-c) show three selected areas **1**, **2** and **3**. The enhanced image are plotted with their respective FFT.

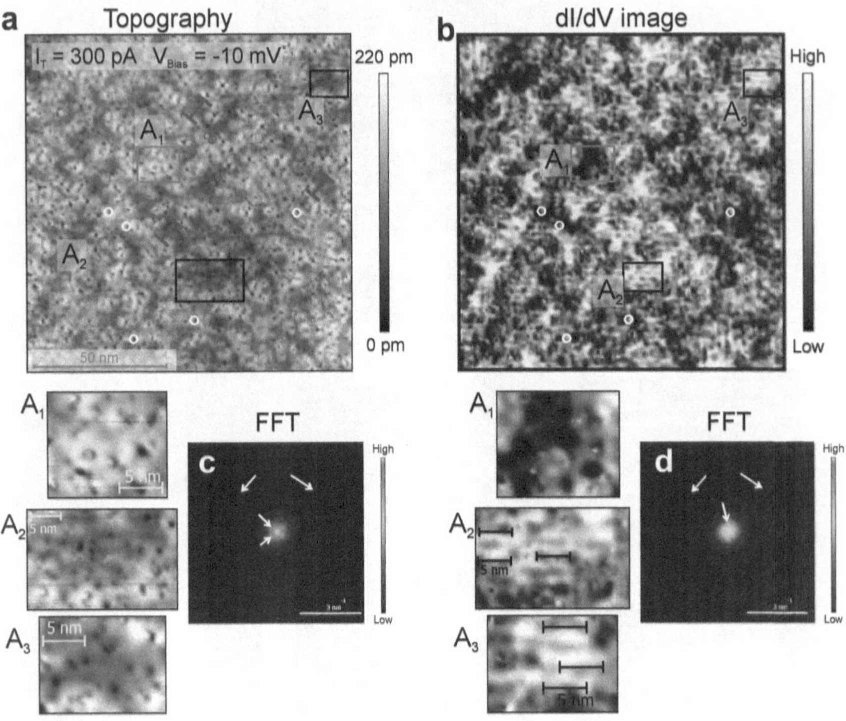

Figure 5.5.6.: 120 nm × 120 nm Topographic and dI/dU images of the atomically resolved surface measured at $T = 5.2$ K with constant tunnelling current
$I_T = 300$ pA and bias voltage $U_{bias} = -10$ mV.

because after the integration the divergence is washed out.

The FFT images for topography (Fig. 5.5.6 (c)) for the dI/dU (Fig. 5.5.6 (d)) show also differences. The FFT of the topography is C$_4$ symmetric. For the FFT of the dI/dU the C$_4$ symmetry is broken by two points in the vertical line crossing the centre.

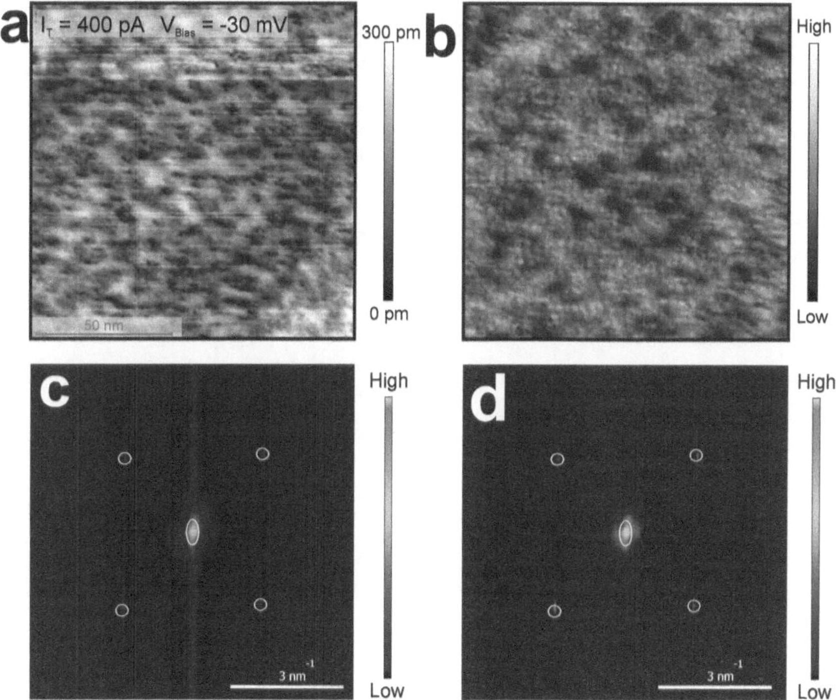

Figure 5.5.7.: 120 nm × 120 nm Topographic and dI/dU images of the atomically resolved surface measured at T = 5.2 K with bias voltage U_{bias} = −30 mV and tunnelling current I_T = 400 pA.

Fig. 5.5.7 shows the surface at the same position as Fig. 5.5.4 and Fig. 5.5.6 measured with U_{bias} = −30 mV. There are big changes with respect to the previous images, at -30 mV in the topography (Fig. 5.5.7 (a)) the sample defects are not distinguishable. Only changes of contrast along large areas of the sample which form clusters. The same situation is seen in the dI/dU image (Fig. 5.5.7 (b)) with the possibility of identifying point structure inside the bright clusters. The FFT images of the topography (Fig. 5.5.7 (c)) and dI/dU image (Fig. 5.5.7 (d)) break the C$_4$ symmetry with a rod-like feature, oriented vertically in the centre of the image.

In conclusion, the topographic and dI/dU images have shown a complex arrangement of the LDOS at the nanoscale. Two different type of defects have been studied at various U_{Bias}. At -10 mV a modulation of the LDOS has been detected with long wavelength (~ 8 nm). Beaking of the C_4 symmetry has been observed in the LDOS by dI/dU and in FFT analysis. Furthermore, the signs of nematic order at 50 mV (Figs. 5.5.4 and 5.5.5) are related to Fe-site defects.

5.5.2. STS of Na$_{0.97}$Li$_{0.03}$FeAs

The surface of Na$_{0.97}$Li$_{0.03}$FeAs shows heterogeneous spectroscopic features at low temperature. Fig. 5.5.8 depicts the three most clear gap among the several spectral features found in the STS data. With a distribution over the sample surface that is not related to clean areas or to crystalline defects. Common to these three spectra is a gap-like shape with different size. The spectra are not fully gapped and show residual DOS at the Fermi level. The size of the largest one, Fig. 5.5.8 (a), is $2\Delta \simeq 40$ meV. Fig. 5.5.8 (b) shows a depletion of intensity near the Fermi level with no coherence peaks. The size of this "pseudogap" is $2\Delta \simeq 20$ meV. Fig. 5.5.8 (c) is the smallest gap, with size $2\Delta \simeq 15$ meV. The gap is highly asymmetric. The amplitude of the peak in the negative is side is much larger than the positive.

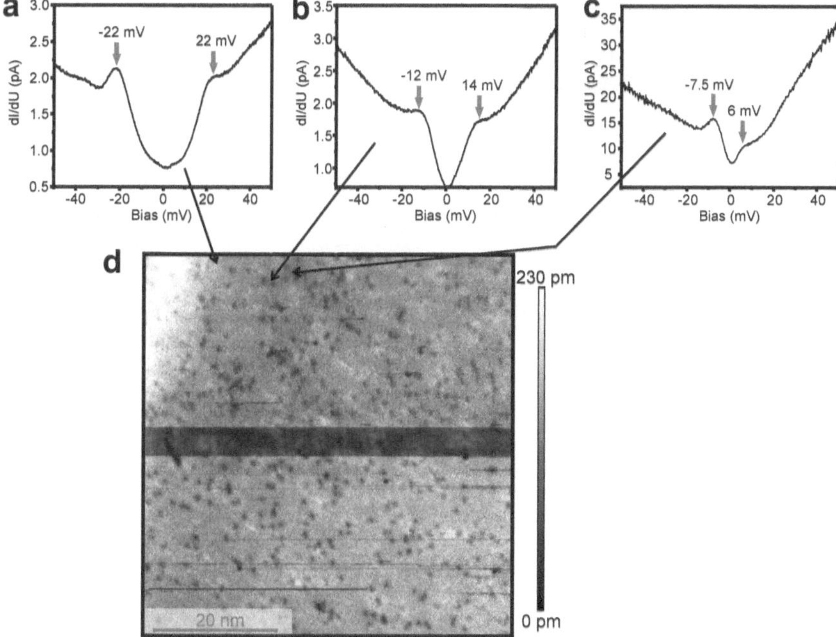

Figure 5.5.8.: Spectroscopic features of Na$_{0.97}$Li$_{0.03}$FeAs measured at $T = 5.2$ K. (a-c) show the three representative spectra observed on the sample surface. (d) Topographic area of 60 nm × 60 nm measured with U$_{Bias}$=100 mV and I$_T$=300 pA corresponding to the are where the spectra (a-c) where taken.

Fig. 5.5.9 (b) shows the dI/dU curves measured along the red line in Fig. 5.5.9 (a). The low temperate spectra show a substantial spatial variation.

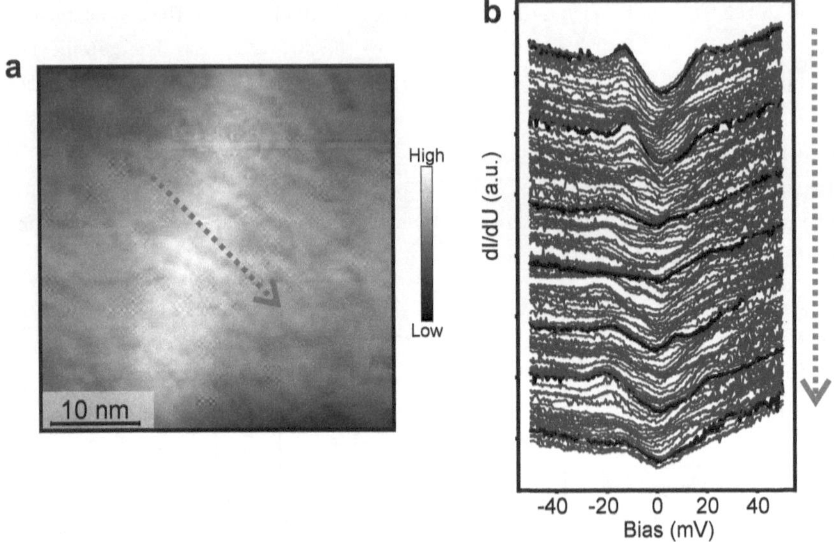

Figure 5.5.9.: Topography corresponding to a conductance map of $Na_{0.97}Li_{0.03}FeAs$ measured in an area of 40 nm × 40 nm at $T = 5.2$ K with $U_{Bias}=50$ mV and $I_T=500$ pA. (d) show the spectra taken along the diagonal red line. Some spectra are highlighted in black to distinguish the spatial variation.

It is possible to compare the data with previous STM studies in the parent compound [33] and Co doped NaFeAs [190, 191] shown in Fig. 5.5.10. The spectrum in Fig. 5.5.8 (a) is similar to the SDW gap observed below the SDW transition in NaFeAs (Fig. 5.5.10 (a and b)). Underdoped $Na(Fe_{1-x}Co_x)As$ samples have proven a strong competition between the SC and SDW gaps [190]. The data (Fig. 5.5.10 (d)) is similar to the line cut shown in Fig. 5.5.9. the situation here is different to optimally doped $Na(Fe_{1-x}Co_x)As$, which shows a homogeneous superconducting gap (Fig. 5.5.10 (c)) not present in $Na_{0.97}Li_{0.03}FeAs$. Nevertheless, the small asymmetric gap in Fig. 5.5.8 (c) may be related to the SC gap. A pseudogap-like spectrum was measured in overdoped $NaFe_{0.939}Co_{0.061}As$ [191]. Differently to the spectrum in Fig. 5.5.8 (c), the pseudogap was asymmetric with respect to the Fermi level. The study assigned the positive bias peak to the superconducting peak and the negative bias pseudogap peak to a magnetic origin. The mechanism has not been further studied. Since the $x = 0.03$ Li doping is in between the two regions in the phase diagram, it is necessary to compare the results with higher doping concentrations.

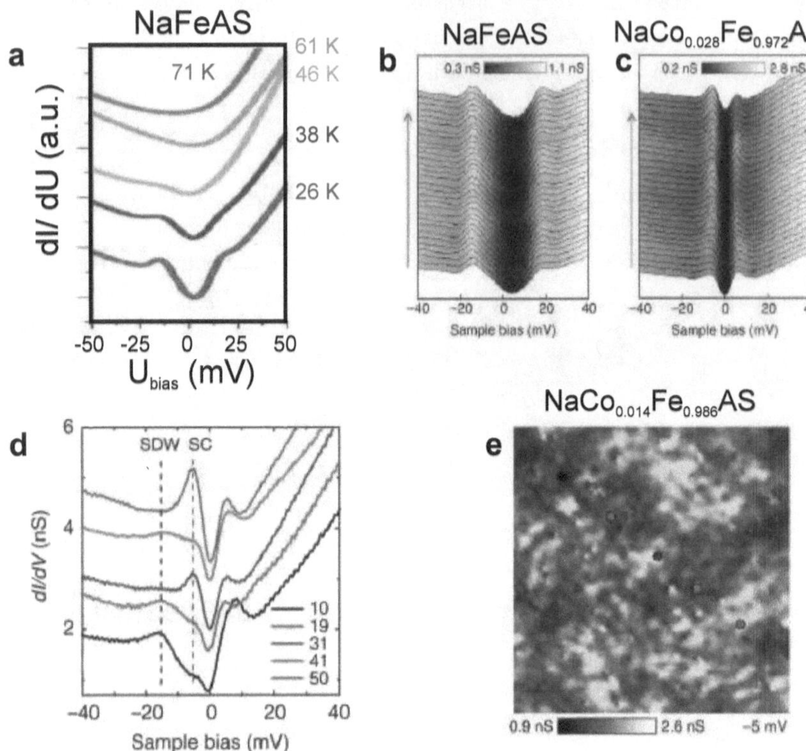

Figure 5.5.10.: STS measurements on NaFeAS and Co doped NaFeAS. (a) shows the temperature evolution of the SDW gap in the STS spectrum of NaFeAs measured in Ref. [33]. (b-c) The dI/dU spectroscopy taken along the red line in Fig 5.5.2 (b-c). The STS spectra from NaFeAs (b) depicts a similar structure as the identified SDW in (a). The results for optimally doped $Na(Fe_{1-x}Co_x)As$ are shown in (c), which depicts a homogeneous superconducting gap. (d) shows dI/dU curves measured in the location marked in (e). (e) conductance maps measured at -5 mV measured on underdoped $Na(Fe_{1-x}Co_x)As$. The STS curves show two distinct gap features, which are more pronounced on the negative side. The negative bump at -17 mV is associated with SDW gap edge while the one at -5 mV is associated with the SC gap edge. The data shows a competition between both orders. The data shown in (b-e) is taken from Ref. [190].

5.5.3. Summary

The thorough surface analysis has displayed a local anisotropy of the tunnelling conductance and its relation to different defects. Primarily, a local breaking of the symmetry C_4 is found for defect D_2 in the dI/dU images. Similar features have been reported in the nematic order phase of the parent compound pinned to Fe defects [33, 160] and in other IBS [32, 163, 164, 168]. A more intriguing observation is found at -10 mV (in Fig. 5.5.6) and -30 mV (in Fig. 5.5.7), where the tunnelling conductance is found to form clusters of electronically ordered patches, which will be addressed in the following sections 5.6 and 5.7.

Interesting features are also observed in the spectroscopy data. It can be related to previous measurement in NaFeAs [33, 190, 191]. The bigger gap shown in Fig. 5.5.8 (a) is reasonably similar to the SDW in the parent compound. However, it is not homogeneously distributed over the surface. This result was expected because previous NMR measurements on the same samples did not probe AFM order. The competition between electronic orders disguises clear signs of superconductivity. The short-ranged AFM order and the "pseudogap" spectrum could be related to the phase transition reported by NMR [199]. To gain more information about the mechanism/s responsible for the gap like features it is necessary to make further studies of the doping and temperature evolution of the spectra.

5.6. Results on Na$_{0.96}$Li$_{0.04}$FeAs

The phase diagram of Na$_{1-x}$Li$_x$FeAs indicates that for 0.04 Li, the SDW has vanished. At this doping below 20 K, the samples develop a newly reported charge/orbital nematic state, where the spin fluctuations are gapped [192]. Here SI-STM was performed at 4.6 K, 12 K and 20 K, in order to study the new nematic state in real space for the first time.

5.6.1. STM/S measurements at 4.6 K

Topography

Fig. 5.6.1 (a) shows a typical topography of Na$_{96}$Li$_{0.04}$FeAs with an area of 60 nm × 60 nm. Measured at 4.6 K. Several types of crystalline defects are present in the surface. The most abundant defects have been enhanced in the inset of fig. 5.6.1 (a). In this image, the atomic corrugation of the Na surface is not distinguishable. Fourier transformation of the topography image (Fig. 5.6.1 (b)) shows the corresponding Bragg peaks of the crystal lattice. Fig. 5.6.1 (c) shows a sketch of the lattice orientation. Additionally, it is possible to observe dumbbell-like defects (see the inset of Fig. 5.6.1 (a)). These dumbbell-like defects are oriented in the direction of the to As-As axes (as seen in the previous topographies of Na$_{0.97}$Li$_{0.03}$FeAs). This orientation is corroborated by the orientation of the Bragg peak in the FFT.

SI-STM

The data collected on Na$_{0.97}$Li$_{0.03}$FeAs showed an intricate spatial variation of the LDOS and STS spectra at the nanoscale. This motivated SI-STM performed on Na$_{96}$Li$_{0.04}$FeAs to achieve a profound investigation of the LDOS. To achieve high-resolution spectroscopy images were measured in an area of 60 nm × 60 nm with 256 px × 256 px for energies between -50 meV to 50 meV, at T = 4.6 K in the same field of view of Fig. 5.6.1.

Fig. 5.6.2 (a-f) portrays selected energies from the measured conductance map. The conductance maps reveal a complex evolution. At -50 meV, in Fig. 5.6.2 (a), two features can be distinguished: i) a majority of point-like bright spots. The bright defects visible in the image account for about 1% of the Na atoms. It is possible that the counting is underestimating the number of defects (because of possible defects in the bright areas of the image which are not accounted). Nevertheless, the estimation is too low to be assigned to Li doping. ii) The LDOS is not uniformly distributed; on the contrary, it is moulded in filaments and clusters (green arrows in the image). In the conductance map at -20 meV (Fig. 5.6.2 (b)) the clusters are better defined. A new observation is discernible in the dark areas. There the LDOS is aligned vertically and horizontally along the Fe-Fe

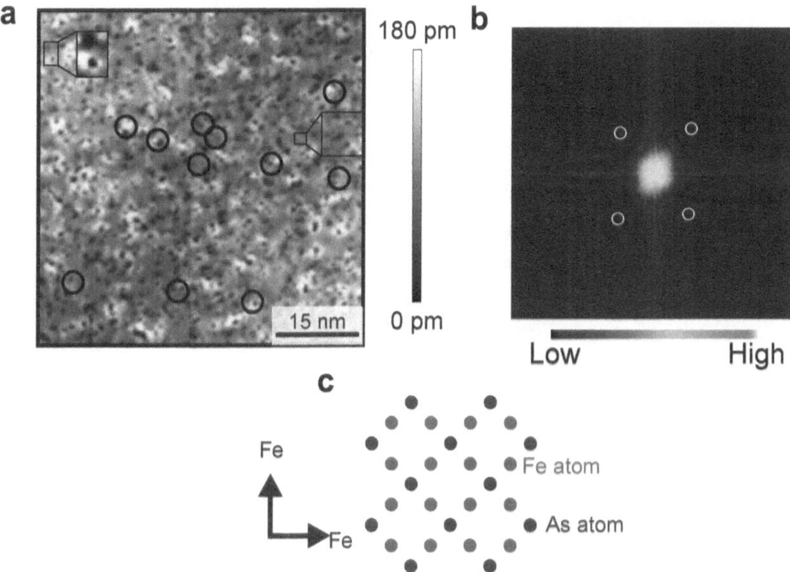

Figure 5.6.1.: (a) Topographic image of the sample surface area of 60 nm × 60 nm, measured at bias voltage $U_{\text{bias}} = 1.0$ V and tunnelling current $I_T = 200$ pA measured at $T = 4.6$ K with 256 px × 256 px. The insets enhance the typical defects, and the dark circles mark the position of nematic defects in Fig. 5.6.2. (b) shows the Fourier transform of (a). The corresponding Bragg peaks of the crystal lattice in concordance with the dumbbell orientation. (c) depicts a sketch of the top view of crystalline structure.

direction. These lines originate a grid-like pattern (it is sketched with dark lines at the centre bottom of the image). At -7.5 meV (in Fig. 5.6.2 (c)) the LDOS form quasi-1D stripe-like patterns along the horizontal Fe-Fe axis, which is now dominant. The image resembles the lines observed in the dI/dU image of Na$_{0.97}$Li$_{0.03}$FeAs measured at -10 meV (Fig. 5.5.6). The distribution of bright and dark areas is inverted from Fig. 5.6.2 (b). Fig. 5.6.2 (d) shows the tunnelling conductance at 7.5 meV. The LDOS is modulated in long-range ($>$ 5nm) connecting the clusters seen in previous energies. It is possible to measure a periodicity of $\sim$ 8 nm marked in the image with red lines. Furthermore, the image show defect with a C$_2$ symmetry feature aligned along the vertical Fe-Fe axis (some examples are marked with dark circles). The nematic defects coincide with some of the bright defects from Fig. 5.6.2 (a and b). For 20 meV, the variance of the conductance is minimum, as it can be seen in Fig. 5.6.2 (d). At 50 meV, the LDOS shows short distance C$_2$ symmetry features like the nematic defects measured in Fig. 5.5.5 for Na$_{0.97}$Li$_{0.03}$FeAs. The bright and dark areas of the conductance present the same modulation of $\sim$ 8 nm with periodicity inverted with respect to Fig. 5.6.2 (d).

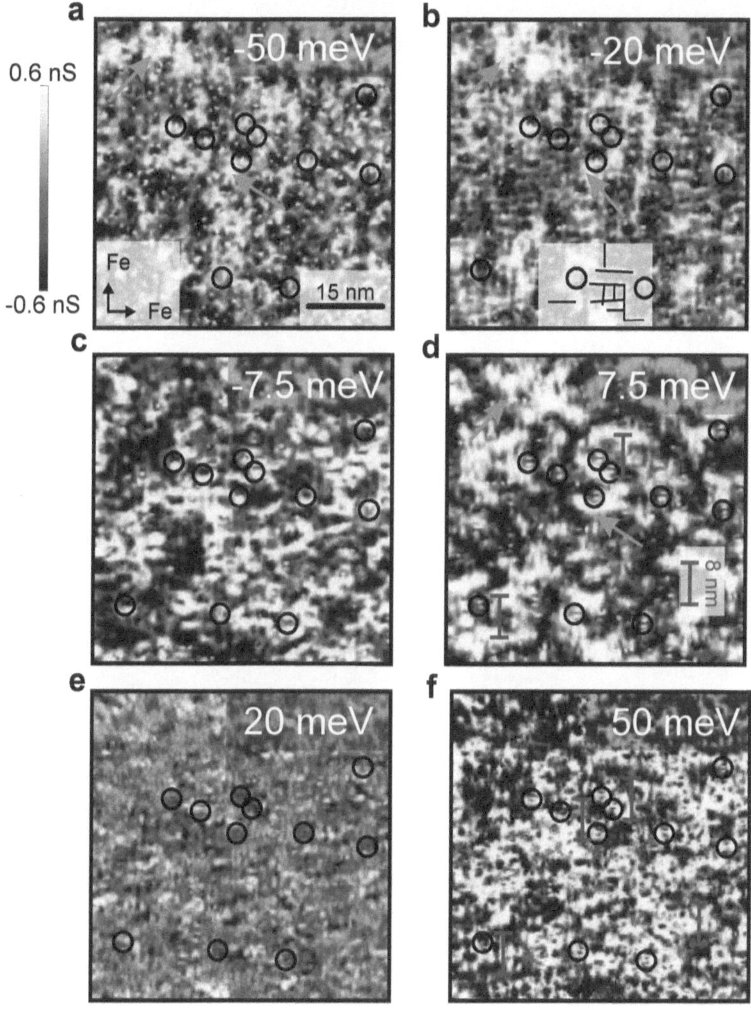

Figure 5.6.2.: (a, b, c, d, e, f) portray the conductance maps for energies between -50 meV to 50 meV. The conductance maps were measured in the same area of Fig. 5.6.1 with 60 nm × 60 nm surface at $T = 4.6$ K. The tunnelling conductance was measured with a lock-in technique with U_{mod}=1.6 mV and setpoint $I_T = 500$ pA and $U_{Bias} = 50$ mV. All the images are plotted after the subtraction of the DC contribution.

Fourier analysis

The two-dimensional FFT is a tool to measure periodicities in a 2D image and therefore uncover the momenta of quasiparticles involved in the scattering processes. Fig. 5.6.3 depicts the FFT analysis correspondent to the conductance maps shown in Fig. 5.6.2. At -50 meV, the intensity is high for small q around (0,0). A horizontal rod centred at (0,0) is distinguishable. It has two smaller satellites parallel in the vertical axis (sketched in dashed lines). Difference plots are used to quantify a possible C_4 symmetry breaking as shown in Ref. [200]. In the difference plot the 2D FFT image is rotated by 90° and subtracted from the original image. If the image is C_4 symmetric, the difference plot is zero at every pixel. Therefore, every non-zero value would quantify the breaking of the C_4 symmetry. In the insets of Fig. 5.6.2 (a) shows broken C_4 symmetry with small intensity. In Fig. 5.6.3 (b) at -20 meV, the features of the FFT change. The extremes of the horizontal rod become isolates points (pointed by the yellow arrow), while a new vertical rod appears (see dashed line). The breaking of the C_4 symmetry as seen in the difference plots with stronger intensity. At -7.5 meV (fig. 5.6.3 (c)) the vertical rod becomes wider and dominant. Hence, the difference plot shows stronger anisotropy, proving nematic order. Interestingly, the two horizontal satellites are less perceptible, however they do not vanish completely. For positive bias energies, the vertical size of the rod shrinks to smaller q. At the same time, the satellites become better distinguishable. As appreciated in the conductance maps, there is a minimum of the FFT signal at 20 meV (fig. 5.6.3 (e)). The signal is maximum at 50 meV (fig. 5.6.3 (f)). For all energies above zero, the C_4 symmetry is strongly broken.

To track the q position of the small circle feature along $(0, \frac{\pi}{a_{Fe}})$ - (0,0) pointed in Fig. 5.6.3 a Gaussian peak is fitted to the FFT at every energy.

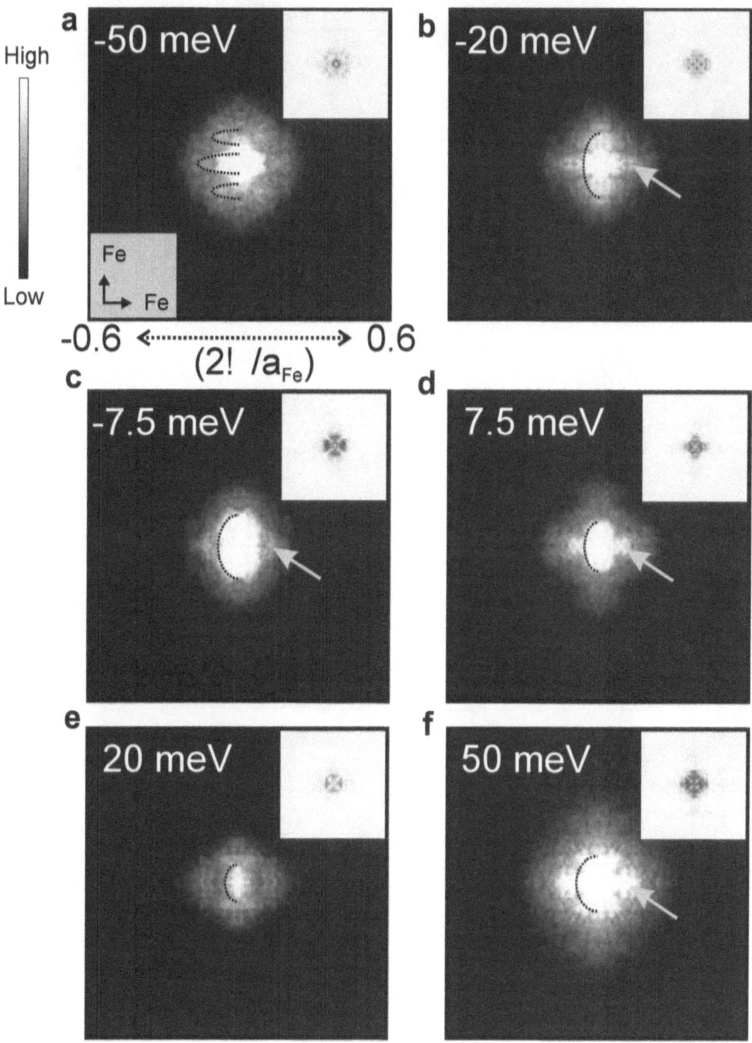

Figure 5.6.3.: The correspondent FFT of the conductance maps of Fig. 5.6.2 for energies between -50 meV to 50 meV. The images are plotted after a symmetrization along x and y direction, a subtraction of a central peak and smoothing. The insets show the difference plots which subtract the FFT image from its rotation by 90°. Difference plots are used to quantify the anisotropy as in Ref. [200].

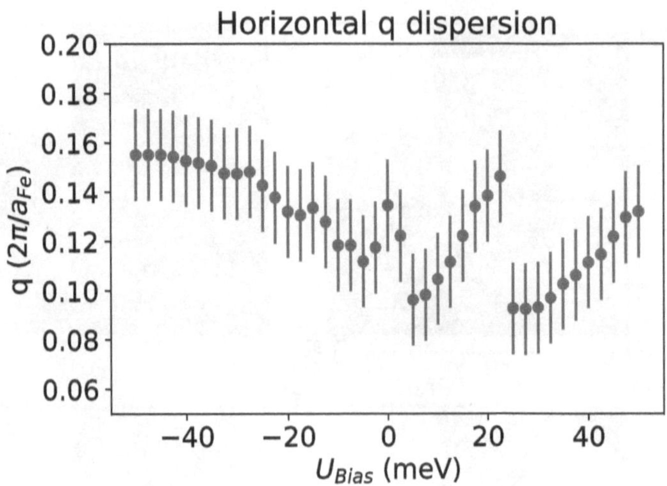

Figure 5.6.4.: q position of a Gaussian peak fitted to the FFT cut along $(\frac{2\pi}{a_{Fe}},0)$ - (0,0) for -30, .27.5, -25 and -22.5 meV.

In Fig. 5.6.4, the extracted q is plotted versus energy.

In general, the Fourier analysis reveals that C_2 symmetry dominates the LDOS. with differences in the QPI scattering along the Fe-Fe directions.

5.6.2. Comparison with previous data

It is beneficial to compare the data with previous results in NaFeAs. This compounds has been used as a benchmark for the detection of nematic fluctuation by QPI. Rosenthal et al. reported local electronic nematicity from the SDW phase up to above the tetragonal phase. Fig. 5.6.5 (a-d) depicts conductance maps of NaFeAs measured in the SDW phase. The first observation is a strong energy dependence, which is also true for the maps shown here for Li doped NaFeAs. At -15 meV (b) and more strongly at 15 meV (c) the images show unidirectional features typical from the nematic order in IBS [32, 164, 201]. The respective FFT are plotted in Fig. 5.6.5 (e-h). The images show a rich structure with energy dependence. A key point is a strong signal along a rod centred at (0,0) and two satellites at $q = \pm 0.5(\pi/a_0)$. The data have a good matching with scattering in the Fermi surface for the SDW phase. In contrast to Refs. [32, 201] a static scattering wavevector is not observed. In $Na_{0.96}Li_{0.04}FeAs$ the FFT show a similar central rod with breaks C_4 symmetry.

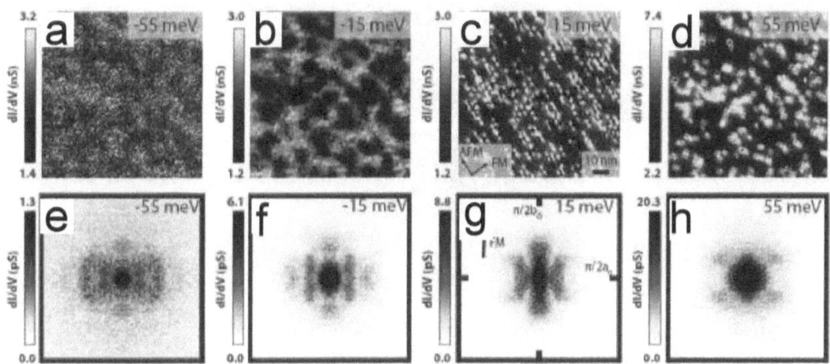

Figure 5.6.5.: (a-d) Differential conductance maps on NaFeAs at 26 K (SDW phase). (e-h) Corresponding FFT images. Data adapted from Ref. [33].

Fig. 5.6.6 shows conductance maps for $Na(Fe_{1-x}Co_x)As$ under- (a-c) and overdoped (e-g) taken from Ref. [160]. The underdoped samples show quasi-unidirectional features (Fig. 5.6.6). The FFT plotted in the insets is like the data shown for $Na_{0.96}Li_{0.04}FeAs$ with positive bias voltage. A significant difference is that the q value for the circular satellites (labelled q_2 in the image) is static with a value of $\approx 0.135(2\pi/a_0)$. For overdoped samples, the nematic fluctuations are insignificant, q_2 shows linear dispersion. It is interesting to point out that the LDOS for -9 meV and - 5 meV resembles the one observed for $Na_{0.96}Li_{0.04}FeAs$ at -7.5 meV. However, the FFT does breaks C_4 in $Na_{0.96}Li_{0.04}FeAs$.

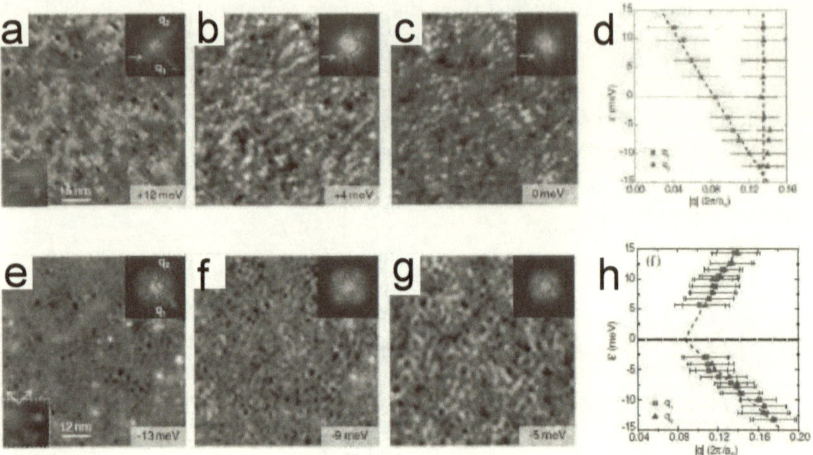

Figure 5.6.6.: (a-c) Differential conductance maps on underdoped $Na(Fe_{0.986}Co_{0.014})As$ at 5 K (Competition between SDW and SC phase). Corresponding FFT images are plotted in the insets. (d) q dispersion extracted from the FFT images of (a-c) along the high symmetry direction. (e-g) Differential conductance maps on optimally doped $Na(Fe_{0.972}Co_{0.028})As$ at 5 K (SC phase). Corresponding FFT images are plotted in the insets. (d) q dispersion extracted from the FFT images of (a-c) along the high symmetry direction. Data adapted from Ref. [160].

5.6.3. Analysis of the QPI patterns

The analysis of this QPI signal can map the electronic structure, which normally is obtained by direct methods like ARPES.

SI-STM has already proven the electronic structure of NaFeAs with successful results for the SDW and nematic phases. Rosenthal *et al.* [33] showed how the features of the QPI in the SDW phase can be deduced from scattering in the Fermi surface. In Fig. 5.6.7 (b) the intensity enclosed by yellow dotted line correspond to scattering amongst the pocket aligned along the same axis (yellow arrows in Fig. 5.6.7 (a)). The four peaks at the diagonals of the image (dotted circles in magenta) correspond to scattering between the two circular pockets along Γ-X and centre pocket with the two circular pockets along Γ-Y (magenta arrows in Fig. 5.6.7 (a)).

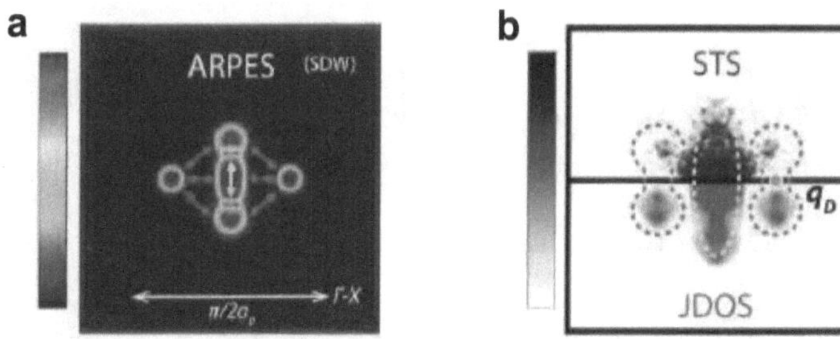

Figure 5.6.7.: (a) ARPES intensity at the Fermi surface for NaFeAs in the SDW phase. (b) Comparison between SI-STM (upper part of the image) and ARPES joint density of states (down part of the image). The dotted lines correspond to scattering peaks produced by the coloured scattering vectors in (a). Data adapted from Ref. [33].

Differently to Fig. 5.6.7 (b) the FFT data for Na$_{96}$Li$_{0.04}$FeAs at 4.8 K (Fig. 5.6.3) only shows two circles along M-Γ $\left(\frac{2\pi}{a_{Fe}},0\right)$. Fig. 5.6.8 displays the extracted energy dispersion of the q from the circle position and the interband scattering between hole-like bands below the Fermi level. The q value extracted from SI-STM has an excellent agreement with the scattering between the α_1 and α_3 bands extracted from ARPES for energies between -50 meV and -10 meV.

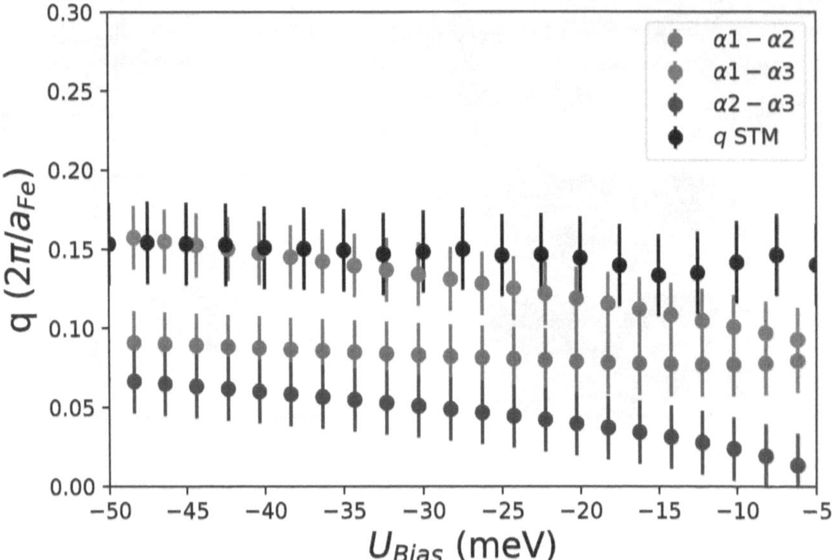

Figure 5.6.8.: Comparison between q vector extracted from STM and interband scattering between the hole-like bands in at the Γ point. In order to obtain the interband scatterings, the data from ARPES (shown in Fig. 5.4.1) was fitted to parabolic dispersions.

5.6.4. STM/S measurements at 12 K

Topography

Fig. 5.6.9 (a and b) depicts two topographies of $Na_{96}Li_{0.04}FeAs$ measured with $I_T = 200pA$ and $U_{Bias} = 100$ mV, at $T=12$ K close to the superconducting transition $T_c \simeq 11$ K. The surface presents similar crystalline defects than the ones measured at low temperature. The Bragg peaks are shown with FFT in Fig. 5.6.9 (c), and they prove that there are no changes with the condition of the SI-STM at 4.8 K. Therefore, the orientation of the lattice is the same.

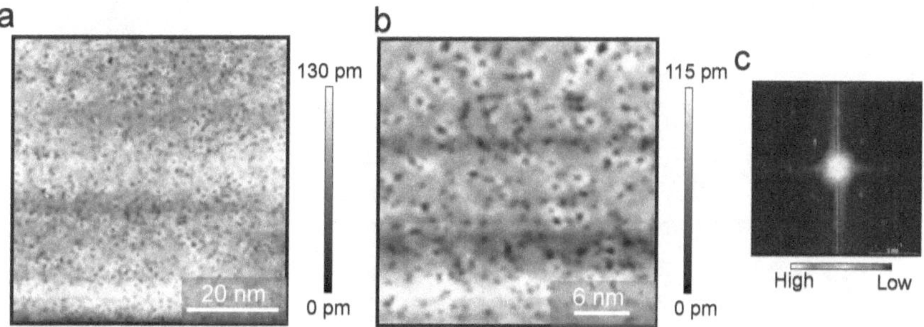

Figure 5.6.9.: Topographic images of the atomically resolved clean surfaces measured at $T = 12$ K. (a) 60 nm × 60 nm topography of the sample surface measured at bias voltage $U_{bias} = 0.1$ V and tunnelling current $I_T = 200$ pA. (b) 30 nm × 30 nm topography measured at bias voltage $U_{bias} = 0.1$ V and tunnelling current $I_T = 200$ pA. (c) shows the Fourier transform of (b).

SI-STM

The SI-STS measurements shown in Fig. 5.6.10 were performed in the area of the surface displayed in Fig. 5.6.9 (b) and in the same sample than Fig. 5.6.2.

Fig. 5.6.10 (a-f) portrays the measured conductance maps for energies between -50 meV to 50 meV, measured in an area of 30 n × 30 nm with 128 px × 128 px at 12 K. In general, the obtained data is very similar to the data previously measured at 4.8 K.

For -50 meV (Fig. 5.6.10 (a)) again, the bright point-like defects and clusterization of the LDOS are distinguishable. At -20 meV (Fig. 5.6.10 (b)) the clusterization is more obvious, and the LDOS is aligned along the Fe-Fe directions with the grid-like pattern

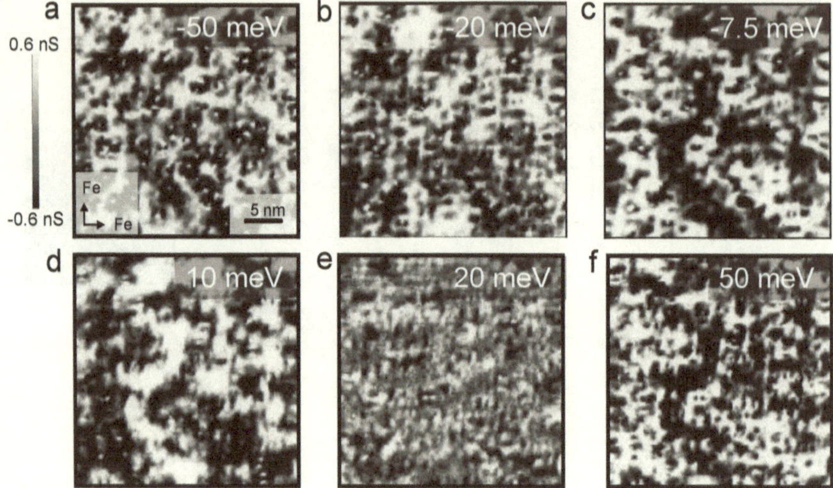

Figure 5.6.10.: (a, b, c, d, e, f) portray the conductance maps for energies between -50 meV to 50 meV. The data was obtained in the same area of Fig. 5.6.9 (b) with 30 nm × 30 nm surface at T = 12 K. The tunnelling conductance was measured with a lock-in technique with U_{mod}=1.6 mV and setpoint $I_T = 500\,pA$ and $U_{Bias} = 50\,mV$. All the images are plotted after the subtraction of the DC contribution.

previously seen. The clusters become dark at -10 meV (Fig. 5.6.10 (c)). The main features of the LDOS are short-range horizontal lines. In Fig. 5.6.10 (d) at 10 meV, the dark and bright contrast is inverted with respect to -7.5 meV. The image area is smaller than the previous data at a lower temperature. Nevertheless, it is still possible to distinguish longer-range modulation of the LDOS with periodicity $\sim$ 8 nm. As seen before for 20 meV the differences in the LDOS intensity are minimized (Fig. 5.6.10 (e)). The differences become stronger at 50 meV with short distance C_2 symmetry features (Fig. 5.6.10 (f)).

Fourier analysis

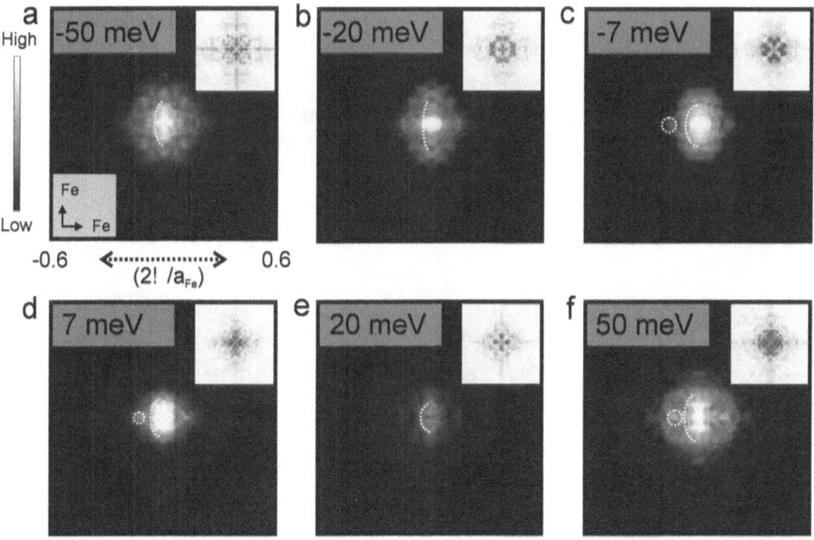

Figure 5.6.11.: The correspondent FFT of the conductance maps of Fig. 5.6.10 for energies between -50 meV to 50 meV. The images are plotted after a symmetrization along x and y direction, subtraction of a central peak and smoothing. The insets show the difference plots used to quantify the anisotropy.

Fig. 5.6.11 depicts the correspondent FFT of the conductance maps from Fig. 5.6.10. At -50 meV, the predominant feature of the FFT intensity is a vertical rod centred at (0,0) (sketched in dashed lines) (Fig. 5.6.11 (a)). This is different than in the data obtained at 4.8 K. However, the difference plot shows a similar breaking of the C_4 symmetry. In

Fig. 5.6.11 (b) at -20 meV, the vertical rod remains present. In the centre along the horizontal axis, two circular satellites appear. The difference plots display a breaking of the C_4 symmetry. For -7.5 meV (fig. 5.6.11 (c)), the vertical rod becomes more intense. Hence, the difference plot shows stronger anisotropy. As seen previously, the vertical size of the rod becomes smaller in q for positive energies. The position of the satellites is only distinguishable at 50 meV. Again, the FFT intensity at 20 meV is minimum, fig. 5.6.11 (e). It becomes stronger again at 50 meV, fig. 5.6.11 (f).

5.6.5. Temperature evolution of the STS data

The STS shows interesting features that evolve through temperature changes from 20 K down to 4.8 K

Fig. 5.6.12 display the spectroscopic data of $Na_{0.96}Li_{0.04}FeAs$ measured from the base temperature at 4.8 K up to 20 K. Fig. 5.6.12 (a-c) show the areas where the conductance maps where measured. An orange dashed line indicates the paths of the plotted spectra in Fig. 5.6.12 (d-f). Several distinct spectra are highlighted in bold black lines to illustrate the vast spectral inhomogeneity. At 4.6 K (Fig. 5.6.12 (d)), the spectra are highly anisotropic at both sides of the bias voltage. A peak-like feature stands out in the negative bias side. The peak position varies from $\simeq -6$ meV to $\simeq -20$ meV. At 12 K (Fig. 5.6.12 (e)), there are no appreciable changes. At T=20 K (Fig. 5.6.12 (f)), the spectra are more homogeneous and, it is difficult to find structures. Is important to note that overall the 20 K data show more noise. Furthermore, the thermal broadening at this temperature is larger, ~ 6 meV at 20 K compared to ≥ 2 meV at 4.6 K, which endanger fine analysis at high temperature.

Although bulk superconductivity is measured in resistivity and magnetization [192], a superconducting gap is not clearly seen in STS. At 4.8 and 20 K, the largest gap has peaks at $\approx \pm 20$ mV, which is similar to the previously measured on $Na_{0.97}Li_{0.03}FeAs$ and resembles the CDW in the literature of NaFeAs [33, 190, 191].

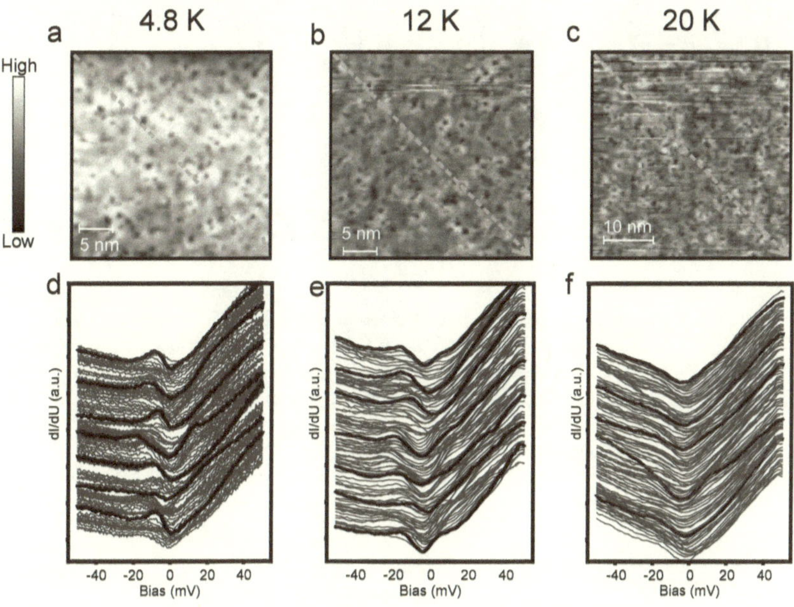

Figure 5.6.12.: Spectroscopic data of Na$_{0.96}$Li$_{0.04}$FeAs measured at 4.6 K, 12 K and 20 K. (a) 60 nm × 60 nm topography of the conductance maps measured at T=4.6 K with a lock-in technique with V$_{mod}$=1.6 mV and setpoint I_T = 500 pA and U_{Bias} = 50 mV. (b) 30 nm × 30 nm topography of the conductance maps measured at T=12 K with a lock-in technique with V$_{mod}$=1.6 mV and setpoint I_T = 500 pA and U_{Bias} = 50 mV. (a) 40 nm × 40 nm topography of the conductance maps measured at T=20 K with a lock-in technique with V$_{mod}$=1.6 mV and setpoint I_T = 400 pA and U_{Bias} = 50 mV. (d-f) Waterfall representation of the spectra taken along the diagonal (orange dashed line) for the topographies (a-c), respectively. Several spectra are coloured in black to highlight the different features found at each temperature.

Δ -map

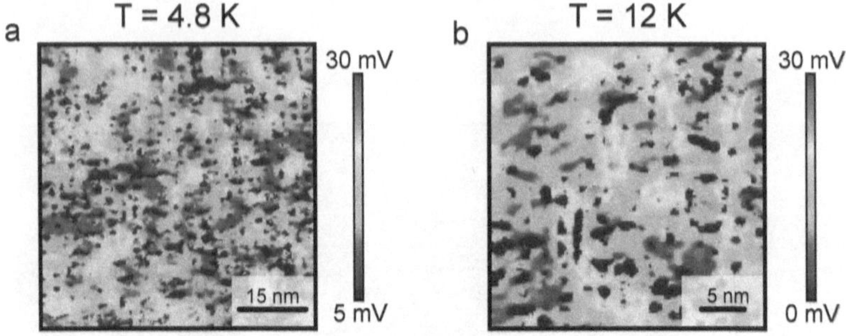

Figure 5.6.13.: Gap maps of $Na_{0.96}Li_{0.04}FeAs$ measured at 4.6 K and 12 K. The gap map are created by fitting a Gaussian peak and a linear dispersion to the negative bias side of the spectra. This approach is used to obtain Δ. (a) Gap map of the 60 nm $\times$ 60 nm conductance map measured at T=4.6 K. (b) 30 nm $\times$ 30 nm topography of the conductance maps measured at T=12 K showed in Fig. 5.6.2. (a) Gap map of the 40 nm $\times$ 40 nm topography of the conductance maps measured at T=20 K shown in Fig. 5.6.10.

Fig. 5.6.13 plot the fitted value of the peak in the negative bias position for the conductance maps measured at 4.6 K and 12 K. Such a map can be denoted as Δ-map or gap-map. It represents the local variation of the spectra of $Na_{0.96}Li_{0.04}FeAs$. The Δ-map strongly resembles the conductance maps measured at $\approx$ 7.5 mV. The Δ-maps show high electronic anisotropy with a modulation of the Δ size similar to the modulation of the LDOS.

K-means clustering

The electronic structure of $Na_{1-x}Li_{x}FeAs$ is strongly inhomogeneous at the nanoscale. An advanced statistical method for automatic separation is K-clustering. The algorithm divides the dataset of N $\times$ N spectra taken in the conductance maps in the selected number of clusters of spectra that have similar behaviour. The algorithm minimizes the distant (sum of squares) of each spectrum within the cluster [202–204]. The centre of each cluster is called centroid and represent the mean value.

$$argmin \sum_{i=1}^{k} \sum_{x_j \epsilon S_i} ||x_j - \mu_i||^2 \tag{5.6.1}$$

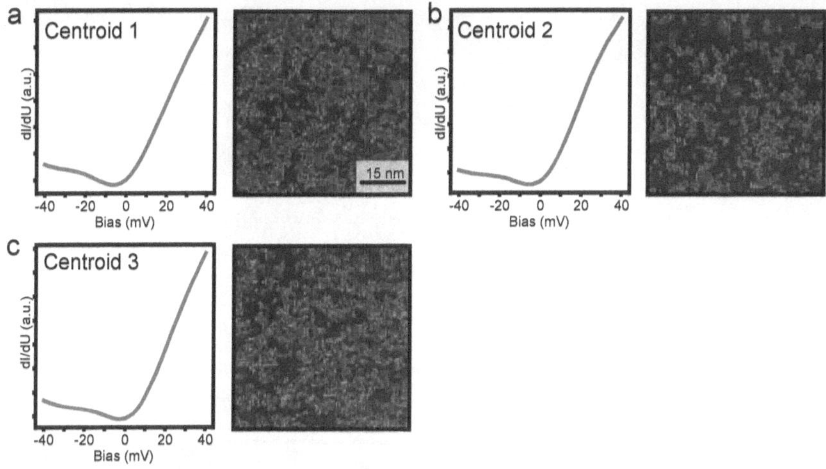

Figure 5.6.14.: K-means clusterization of $Na_{0.96}Li_{0.04}FeAs$ measured at 4.6 K in the area shown in Fig. 5.6.2. (a-c) represent the mean spectrum (centroid) and its spatial distribution for k=3 clusters.

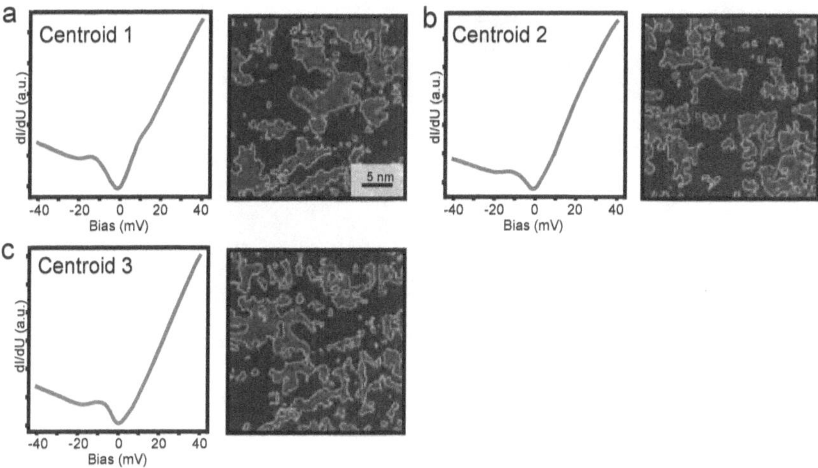

Figure 5.6.15.: K-means clusterization of $Na_{0.96}Li_{0.04}FeAs$ measured at 4.6 K in a zoom of 30 nm × 30 nm of the area shown in Fig. 5.6.2. (a-c) represent the mean spectrum (centroid) and its spatial distribution for k=3 clusters.

For the data in this **book**, the k-cluster maps were generated using the Python library scikit-learn [205]. Figs. 5.6.14 and 5.6.15 represent the clusterization of conductance maps measured at 4.6 K. The data sets consist of 256 pixel × 256 pixels × 41 energy channels in a 60 nm × 60 nm area (Fig. 5.6.14) and 128 pixels × 128 pixels × 201 energy channels within an area of 30 nm × 30 nm in a zoom-in area of Fig. 5.6.14 (Fig. 5.6.15). The clusterization of conductance maps measured 12 K is depicted in Fig. 5.6.16 for an area of 30 nm × 30 nm, with 128 pixels × 128 pixels × 101 energy channels.

Centroid 1 in Figs. 5.6.14, 5.6.15 and 5.6.16 shows a similar spatial distribution to the electronic modulation present in the conductance maps and the gap-maps. The spectra for centroid 1 is not particle-hole symmetric and have two peaks at ≈-15 mV and ≈10 mV. The spatial distribution of the Centroids 3 can be related to the grid-like ordering of the LDOS measured between -20 and -15 mV.

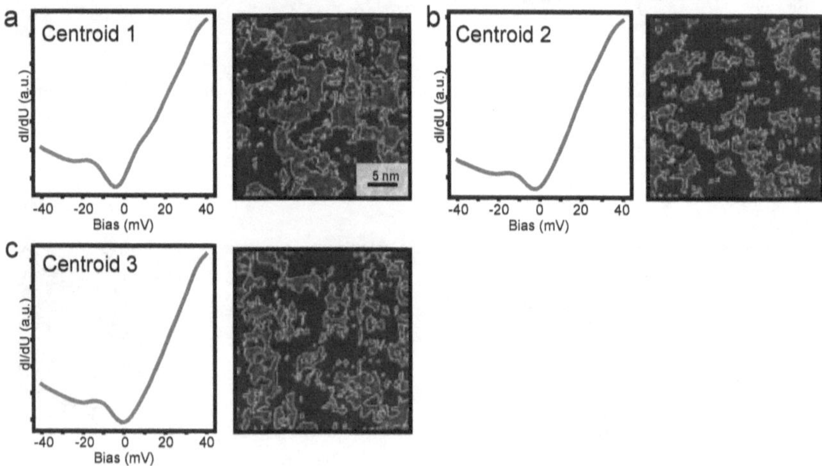

Figure 5.6.16.: K-means clusterization of Na$_{0.96}$Li$_{0.04}$FeAs measured at 12 K. In the area shown in Fig. 5.6.10. (a-c) represent the mean spectrum (centroid) and its spatial distribution for k=3 clusters

5.6.6. Summary

The main phenomena observed in this section are:

1. Point-like atomic defects, which show C_2 symmetry in the LDOS.

2. Although the surface possesses a strongly inhomogeneous electronic landscape at the nanoscale. Nevertheless, a modulation in the electronic structure with periodicity $\sim$ 8 nm have been observed in the SI-STM data at 4.8, and 12 K.

3. Nematic order has been revealed by the FFT study of the conductance maps, with strong energy dependence.

4. A q value extracted from FFT for energies between -50 meV and -10 meV shows QPI scattering, which is in agreement with the scattering between the α_1 and α_3 bands extracted from ARPES

5. Gap-maps and k-means clusters present similar order. The comparison hints to a common mechanism for the ordering of the LDOS in real space and the broad range STS spectra.

5.7. Search for CDW on Na$_{0.96}$Li$_{0.04}$FeAs

The conductance and gap maps in Na$_{0.96}$Li$_{0.04}$FeAs have shown a modulation in the electronic structure. This modulation is proven to influence the LDOS and the STS spectra. These observations are particularly interesting since the modulation is static. It could be related to the charge/orbital state observed in NMR [192]. The previous sections have encouraged the search for charge instabilities in the electronic structure. Here the energy dependence of the LDOS modulations is studied with q-selective inverse-Fourier transform analysis and q-selective spectroscopy. Additionally, phase-sensitive STS is used to relate those changes to the dI/dU.

5.7.1. Energy dependence

Starting with a characterization of the LDOS modulation and its energy dependence. Fig. 5.7.1 displays four conductance maps at different energies and the inverse-Fourier transform image (IFTi). The FFT images have been filtered for small circle region around (0,0) with a radius of $q \sim 0.08 \frac{\pi}{a_{Fe}}$, which correspond to the modulation observed in Figs. 5.6.2-5.6.10 with periodicity ~ 8 nm. Fig. 5.7.1 (a) depicts the conductance map for 7.5 meV, where the modulation has its intensity maximum. To study the energy dependence, two areas where the electronic modulation is robust are delimited with green squares. The IFTi in Fig. 5.7.1 (b) enhances the main features of electronic modulation. Easiest to distinguish is an S-like shape in the central square. At 35 meV in Fig. 5.7.1 (c), an inversion of the modulation phase can be appreciated. It is easier to appreciate inside the green square and more evident in the IFTi maps (Fig. 5.7.1 (d)), where the S figure is now in dark. Fig. 5.7.1 (e) and (f) depicts the conductance and IFTi map for -7.5 meV, respectively. The modulation is not the dominant feature, there is high contribution form QPI. However, resemblances with the conductance map at 35 meV are possible to recognize. At this energy, the horizontal electronic scattering is very strong and interferes with the static modulation. At -20 mV the conductance map (Fig. 5.7.1 (g)) shows strong electronic scattering with a grid pattern, yet the electronic modulation is recognizable in the IFTi (Fig. 5.7.1 (h)). Here the q-selective IFTi analysis has shown that the electronic modulation observed in the conductance maps has no energy dispersion.

The same q-selective IFTi analysis is performed in the 12 K SI-STM data. The conductance maps and IFTi images for energies 35, 7, -7 and -20 meV are plotted in Fig. 5.7.2 (a-h). An S-like shape is highlighted inside a green square for Fig. 5.7.2 (a and b) at 7 meV, to confirm the changes in the phase of the CDW modulation. This S-like feature is dark for the bias energies 35 and -7 mV. Again, bright at -20 mV. Although the conductance maps are not as large as the ones measured at 4.8 K, similar static modulation of

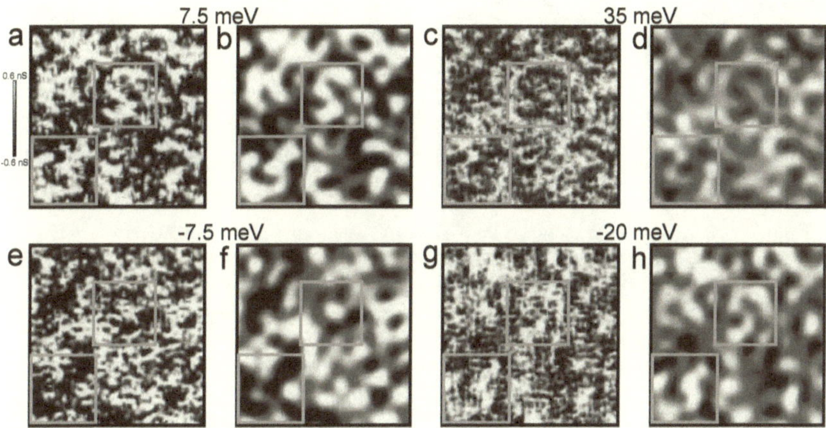

Figure 5.7.1.: (a) Conductance map at 7.5 mV. The green squares delineate areas to enlighten the long-range modulation. The circles point the defect with C_2 symmetry, vertical (in green) and horizontal (white). (b) Inverse-FFT (IFFT) image of (a) for $q \sim 0.08 \frac{\pi}{a_{Fe}}$. (c) Conductance map at 35 mV. The bright long-range modulation is now minimum in the green squares. There is a phase inversion for the CDW as can be appreciated in the IFFT image (d). (e) shows the conductance map at -7.5 mV. The same kind of sign inversion is visible for the areas inside the squares. This sign inversion now is more clear for the IFFT image in (f). (g) Conductance map at -20 mV. (h) show the IFFT image with dame phase as in (b).

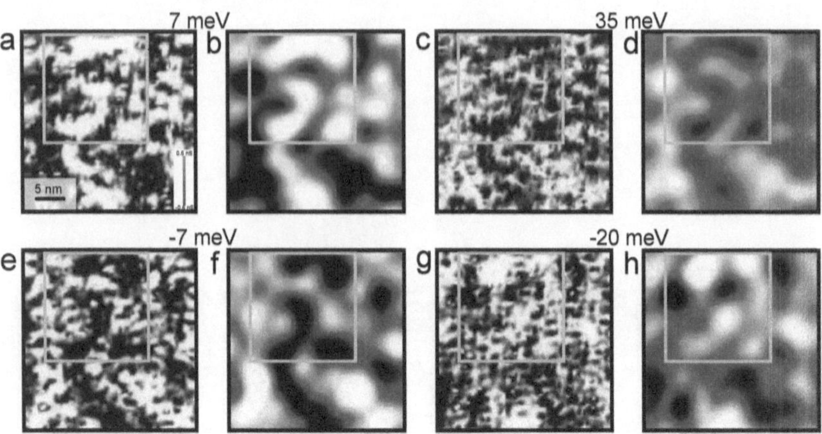

Figure 5.7.2.: (a) Conductance map at 7 mV. A green square enlightens the long-range modulation.   (b) Inverse-FFT (IFFT) image of (a) for $q \sim 0.08\frac{\pi}{a_{Fe}}$.   (c) Conductance map at 35 mV. The long-range modulation is now dark in the green square. The phase inversion for the CDW as can be appreciated in the IFTi (d). (e) shows the conductance map at -7 mV. Inside the green square, the modulation is dark. The phase is more visible for the IFTi in (f). (g) Conductance map at -20 mV. (h) show the IFTi with dame phase as in (a and b).

the LDOS is observed at 12 K.

The presence of a non-dispersive electronic modulation at 4.8 and 12 K strengthens observation of charge order in $Na_{0.96}Li_{0.04}FeAs$.

5.7.2. Electronic structure

The data extracted from SI-STS facilitates the search of a plausible charge instability in the electronic structure. The measured modulation has a wavelength of $q \sim 0.08\frac{\pi}{a_{Fe}}$. Therefore a nesting vector with similar k is needed. The ARPES data (Fig. 5.4.1) showed a small hole-pocket crossing the Fermi level at the Γ point with $q_{2k_F} \sim 0.05\frac{\pi}{a_{Fe}}$, which is in the range of the number obtained by our STS measurements.

More work is done on the STS data. The FFT intensity is integrated inside the same region used for the IFFT analysis ($q \sim 0.08\frac{\pi}{a_{Fe}}$) looking for more pieces of evidence of the charge instability. The q-selective spectroscopy is shown in Fig. 5.7.3 (a) shows for all energies at 4.8 K. q-selective spectroscopy reveals a gap opening with size $2\Delta = 15$ meV.

Fig. 5.7.4 displays the q-selective spectroscopy measured at 12 K. The data confirm

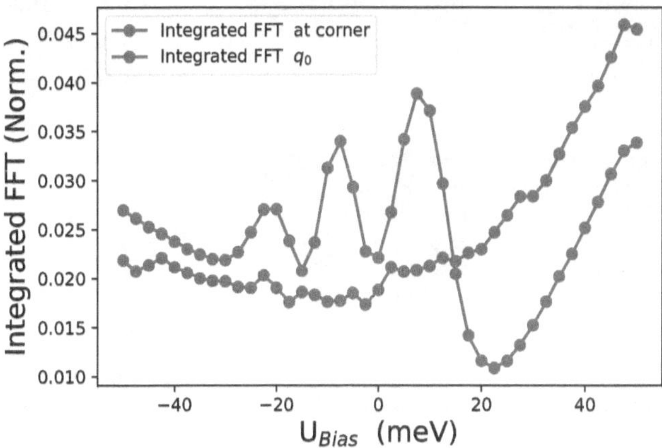

Figure 5.7.3.: Q-selective spectroscopy at the Γ point and at $(\frac{\pi}{a_{Fe}}, \frac{\pi}{a_{Fe}})$. The figure shows integrated FFT intensity in the area $q \sim 0.08\frac{\pi}{a_{Fe}}$ for all energies at 4.8 K.

the presence of a gap with size $2\Delta = 15$ meV.

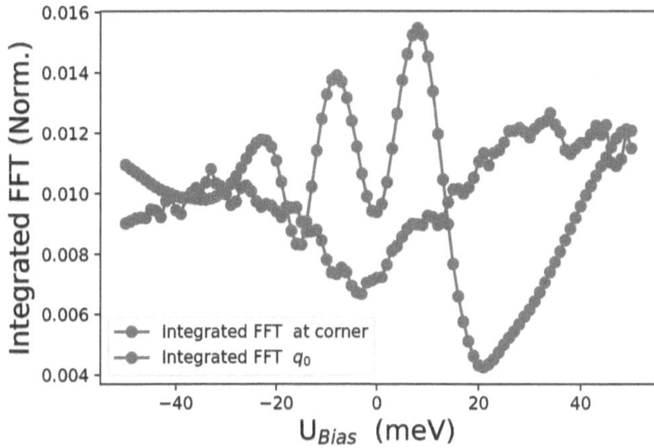

Figure 5.7.4.: Q-selective spectroscopy at the Γ point and at $(\frac{\pi}{a_{Fe}}, \frac{\pi}{a_{Fe}})$. The figure shows integrated FFT intensity in the area $q \sim 0.08\frac{\pi}{a_{Fe}}$ for all energies at 12 K.

The comparison with ARPES pinpoints the found gap to α_3 (the hole-like band at the Γ-point). A third peak is found in the q-selective spectroscopy at -20 meV, which can be explained with the help of the band structure (Fig. 5.7.5 (a)). At -20 meV the top of the α_2 band is present, it naturally enhances the conductance at -20 meV.

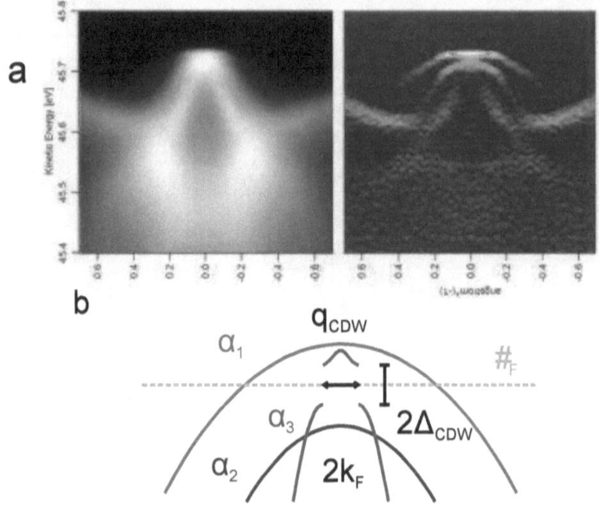

Figure 5.7.5.: (a) Low-temperature ARPES data for horizontal and vertical polarization and its second derivative. The localization in the momentum space of the CDW, $q \sim 0.08\frac{\pi}{a_{Fe}}$, allow us to look for the opening of a gap in ARPES. Where we find a narrow band with d_{yz} character crossing the Fermi level at $q_{2k_F} \sim 0.05\frac{\pi}{a_{Fe}}$. (b) sketches the opening of the CDW gap in the narrowest d_{yz} band at the Γ point. ARPES data courtesy of Y. Kushnirenko, S. Federov and S. V. Borisenko

Here q-selective spectroscopy has identified the opening of a gap and successfully measured its size ($2\Delta = 15$ meV). From the ARPES data the hole-like band α_3 is the most likely host of the gap. The energies coincide with the relative maxima found at 7.5, -7.5 and -20 meV and the minimum at 20 meV from the cuts of the FFT matrices along the high symmetry directions. The opening of the CDW gap is illustrated in Fig. 5.7.5 (a). Furthermore, the inversion of the contrast of the conductance maps pointed in Fig. 5.7.1 can also be explained with a known feature of the CDW. The change of the CDW phase at $\frac{dI}{dU}(E_F + \Delta)$ (7.5 meV) and $\frac{dI}{dU}(E_F - \Delta)$ (-7.5 meV) CDW [206]. This change is sketched in Fig. 5.7.6.

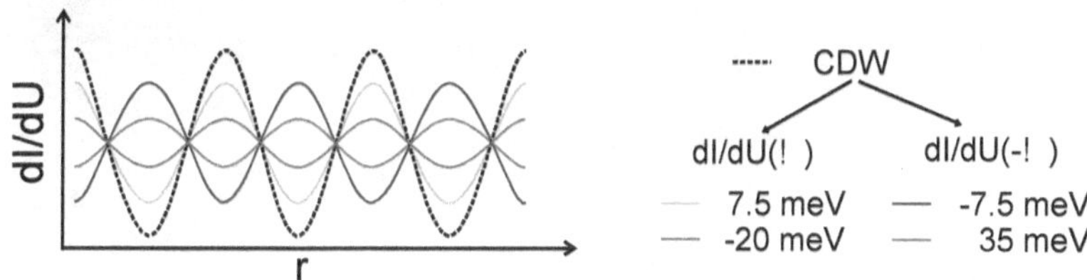

Figure 5.7.6.: Sketch of the CDW and the modulation of the LDOS for different energies.

5.7.3. Phase-selective STS

A phase-selective study is done on the STS data in order to determine the influence of CDW in the tunnelling spectra.

In this study, the conductance map at the value corresponding to $\frac{dI}{dU}(E_F+\Delta)$ (7.5 meV) is divided into two areas: One where the CDW phase is close to its maximum (orange region) and a second where the CDW phase is close to its minimum (purple region). This phase-map is plotted in Fig. 5.7.7 (a), the insets display the average spectrum in both regions.

The difference between maximum and minimum averages is plotted in Fig. 5.7.7 to better illustrate the variance in the spectra. The phase differentiation of the spectroscopic data shows a maximum at +7.5 meV and a minimum at -7.5 meV, which is an expected signature of the CDW phase change $\phi(\Delta) = -\phi(-\Delta)$.

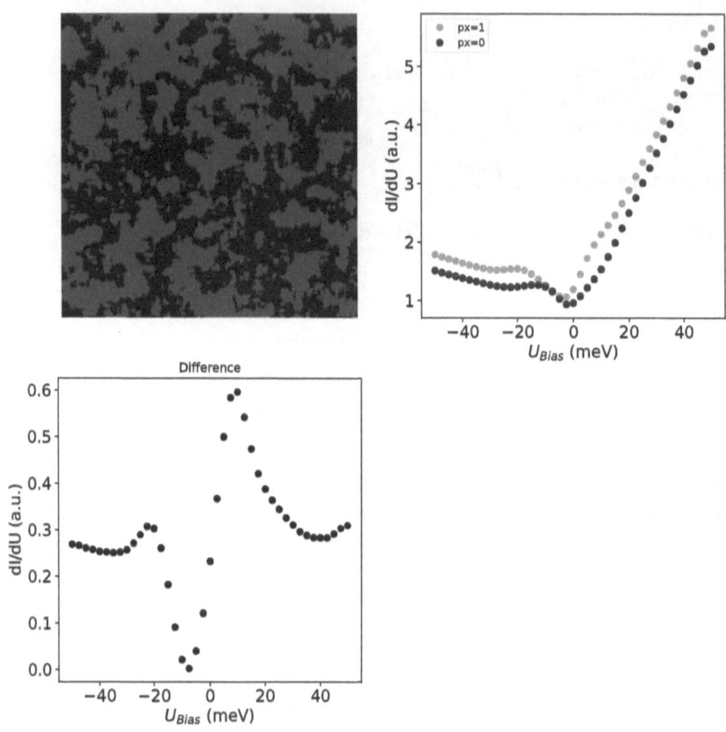

Figure 5.7.7.: Mask map of the conductance at 7.5 meV. For every pixel, if LDOS(x,y) $\geq 0.01 nS$ the value is set to 1, else the pixel is 0. This allow the distinction of two areas in the conductance maps. In one area the amplitude of the CDW is maxima and The second where is minima. The correspond average spectra for both areas, and their difference are displayed.

5.7.4. Summary

This section has successfully identified the presence of a CDW which modulates the LDOS. q-selective spectroscopy has found the CDW gap with $2\Delta_{CDW} = 15\ meV$ and $q \sim 0.08\frac{\pi}{a_{Fe}}$. ARPES data have pinpointed the CDW to the d_{yz} band at the Γ point. Finally, the phase-sensitive spectroscopy has proven the effects of the CDW to the local $\mathrm{d}I/\mathrm{d}U$, and the local variance of these effects.

5.8. Discussion

The first part of this chapter shows the results obtained in $Na_{0.97}Li_{0.03}FeAs$. It has shown nematic order pinned to Fe-site defects. Furthermore, the Fourier analysis confirmed that C_4 symmetry is globally broken. The results are similar to other STM measurements in the nematic phase for NaFeAs [33, 160] or other IBS members [32, 163, 164, 168]. The spectroscopic features are typical of competition among electronic phases. At -10 mV (in Fig. 5.5.7) the tunnelling conductance is found to form clusters of electronically ordered patches. This is very similar to the CDW modulation measured for $Na_{0.96}Li_{0.04}FeAs$. The spectroscopy of $Na_{0.97}Li_{0.03}FeAs$ has revealed different features, which resembles the previous measurement in NaFeAs [33, 190, 191]. The results evidence an electronic competition between ordered phases.

$Na_{0.96}Li_{0.04}FeAs$ has been studied in the second part of the chapter. SI-STM has revealed an electronic modulation of the LDOS with periodicity ~ 8 nm. The Inverse-Fourier transform analysis has shown that the modulation is not dispersive and q-selective spectroscopy has determined the opening of a CDW gap. Moreover, the analysis of the band structure (provided by ARPES) pinpointed the gap opening to the α_3 band at the Γ point. Finally, a phase-selective analysis of the STS has reconciled the Δ and K-means -maps with the effect of CDW in the tunnelling spectra.

The anisotropy of the Fourier analysis of the conductance maps and the C_2 symmetric defects has shown an intimate connection between the CDW and the nematic order. It has been predicted that the orbital character of the band has strong consequences in the nematic instability due to the orbital selectivity of the spin-fluctuation [169]. The scenario of orbital selectivity spin-fluctuations can potentially explain the abrupt reduction of the spin-lattice relaxation rates measured by NMR [192] to the opening of a CDW in the d_{yz} band. Besides, the change in the spin-fluctuations should be responsible for the different nematic order. Different mechanisms have been proposed for the formation of CDW in IBS, like Spin density wave [207] or vestigial order[1] [208]. Previous observations of CDW in IBS are very scarce: FeSe has shown stripes in the FM direction [162, 209]. Additionally, charge instabilities have been reported in Mn-substituted LaFeAsO [210] and the structural homologue $Ba(Ni_{1-x}Co_x)_2As_2$ [211]. These results support the experimental observations done with SI-STS in this book. These findings add a further piece to the puzzle of the phase diagram of IBS and expand the understanding of the relationship between charge order, magnetic order and unconventional superconductivity.

---

[1] for multicomponent order parameter there are situations where the primary order cannot establish long-range order. However the condensation of fluctuations at their own transition temperature is possible, these order phases are called vestigial orders, because they are a remnant of the primary order.

6. Summary

The focus of this experimental work is the investigation of emergent quantum phases in novel correlated electron systems with scanning tunnelling microscopy. Two different systems have been studied: Sr_2IrO_4, a novel kind of Mott insulator. For the $5d$ valence electrons of Ir, the spin-orbit coupling of the valance electrons is comparable to the strength of the correlations. Here the magnetic state emerges from a combination of spin-orbit coupling and electron-electron correlations. The second studied material is $Na_{1-x}Li_xFeAs$, a member of the Iron-based superconductors family. These materials offer a new platform to study unconventional superconductivity (electronically driven) and its relation to several electronic orders. In the IBS, the strength of the electronic correlations is moderate. Nevertheless, they manifest with strong spin an orbital fluctuations.

The results on Sr_2IrO_4 are shown in chapter 4. There the motion of charges in the $J_{eff} = 1/2$ AFM background of Sr_2IrO_4 has been investigated. The STS spectra from clean areas of the sample surface revealed peak- or shoulder-like anomalies above and below the Mott gap. When the spectra are compared with the self-consistent Born approximation of the polaronic model in the AFM background gives excellent quantitative agreement. The anomalies in the unoccupied states have been successfully assigned to the spin-polaron ladder spectrum. Whereas in the occupied region, the charge excitation has internal degrees of freedom and therefore, the movement of the spin-polaron becomes more complex and highly dispersive. Nevertheless, the singlet and triplet polarons are found responsible for the anomalies. These are the first experimental observations of the spin-polaron and these internal degrees of freedom in Mott insulators, which have been later confirmed by cold-atoms lattice simulators [114]. The work here incentives to revisit previous studies, on cuprates and other Mott insulators at half-filling, in the search for more signs of spin-polarons quasiparticles. Additionally, the value of electron-electron correlations, U, has been calculated for Sr_2IrO_4 between 2.05 eV and 2.18 eV.

The second part of the chapter studies the metal-insulator transition (MIT). The SI-STS analysis of the defect effects on Sr_2IrO_4 has revealed a nanoscale region where the conductivity is higher due to in-gap states (defect bound states) with 2 nm lengths. The study of the atomically resolved topographic images has delimited the position of the defect to the centre of the IrO_6 octahedra. In concordance with previous studies [72] apical oxygen vacancies are the most likely candidate. Percolation through such vacancies has been proposed as the mechanism responsible for the MIT. The percolation model gives a 3.7% threshold, which agrees with the experimental critical doping level [72]. Filamentary charge conduction paths have been found in conductance maps and topography. Such scenario is fundamentally different from the usual mechanism of the metal-insulator transition in other Mott insulators (cuprates) originated from charges moving in the correlated AFM background and therefore, questions a global doping scheme for Mott insulators.

The study of $Na_{1-x}Li_xFeAs$ is shown in chapter 5. The first part displays the results obtained for $Na_{0.97}Li_{0.03}FeAs$. The analysis of the topographies and dI/dU maps showed how the Fe-site defect break the electronic C_4 symmetry. The nematic feature is globally measured in Fourier analysis which confirms the breaking of the C_4 symmetry. The study of the dI/dU spectra have shown a very complex panorama where various electronic orders are competing.

The second part of the chapter displays the experimental work performed on $Na_{0.96}Li_{0.04}FeAs$. Here, the SI-STS revealed an electronic modulation of the LDOS with periodicity ~ 8 nm. Inverse-Fourier analysis of the conductance has proven that the modulation is not dispersive. The q-selective spectroscopic analysis has measured the opening of a gap with size $2\delta = 15$ meV. ARPES measurements pinpointed the gap location to the α_3 band at the Γ point. Finally, the long-range modulation has been successfully identified as CDW in the material. The STS spectra have shown a similarly complex scenario to $Na_{0.97}Li_{0.03}FeAs$. It has been analysed with Δ and K-means -maps, in both cases the CDW has been shown responsible for the changes in the tunnelling spectra.

The $Na_{1-x}Li_xFeAs$ investigations have shown the connection between the CDW and the nematic order. Later, the relation has been corroborated by the anisotropy of the Fourier analysis of the conductance maps, and the C_2 symmetric defects. The search of the electronic order measured with NMR [192] has been completed. These results are the more robust determination of charge order in the IBS, although scarce previous results were already known in the literate [162, 209–211]. The results here undercover a new piece of the IBS phase diagram. Moreover, motivate further studies of CDWs in other IBS members.

In conclusion, all the above examples show how the novel quantum phases emerge from the interplay between charge, spin and orbital degrees of freedom.

A. 2-Dimensional Fourier transform

In chapter 5, the images of the 2-dimensional Fourier transformation are plotted after a symmetrization along x and y direction, subtraction of a central peak and smoothing. Here the raw data is showed together with the symmetrized and the symmetrized and smoothed data. Changes in the data are negligible and therefore artefacts are not seen.

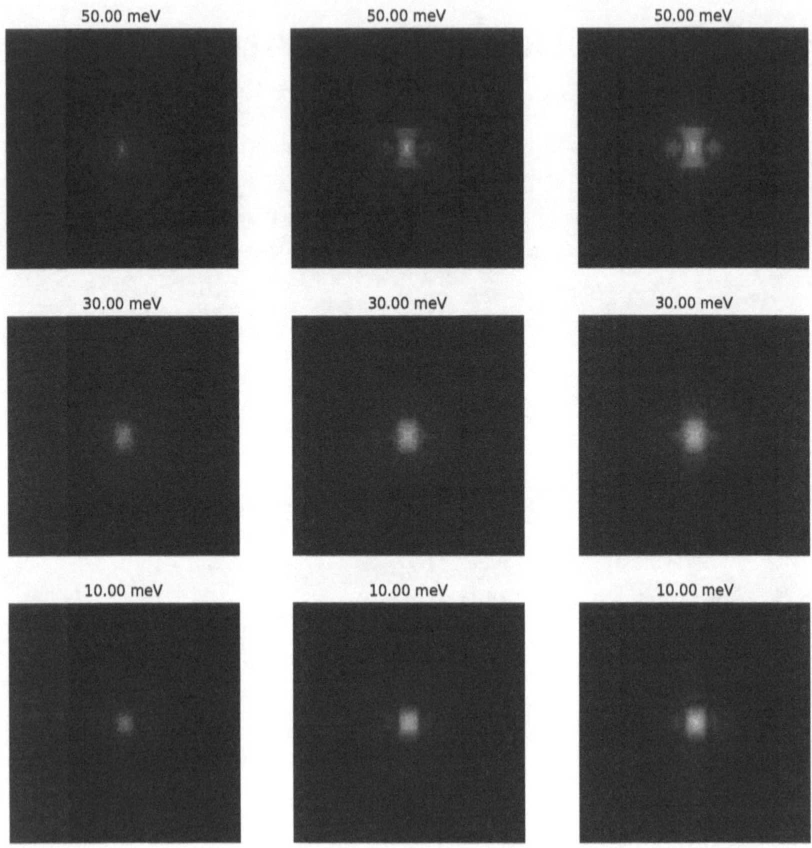

Figure A.0.1.: 2-dimensional FFT data from conductance map n. 11, measured at 4.6 K with U_{mod} =1.6 mV, U_{bias} =50 mV and I_T = 500 pA, in an area of size 60 nm × 60 nm with 256 px × 256 px. Raw data is shown at the left of the image, symmetrized at the centre and smoothed at the right for energies 50, 30 and 10 meV.

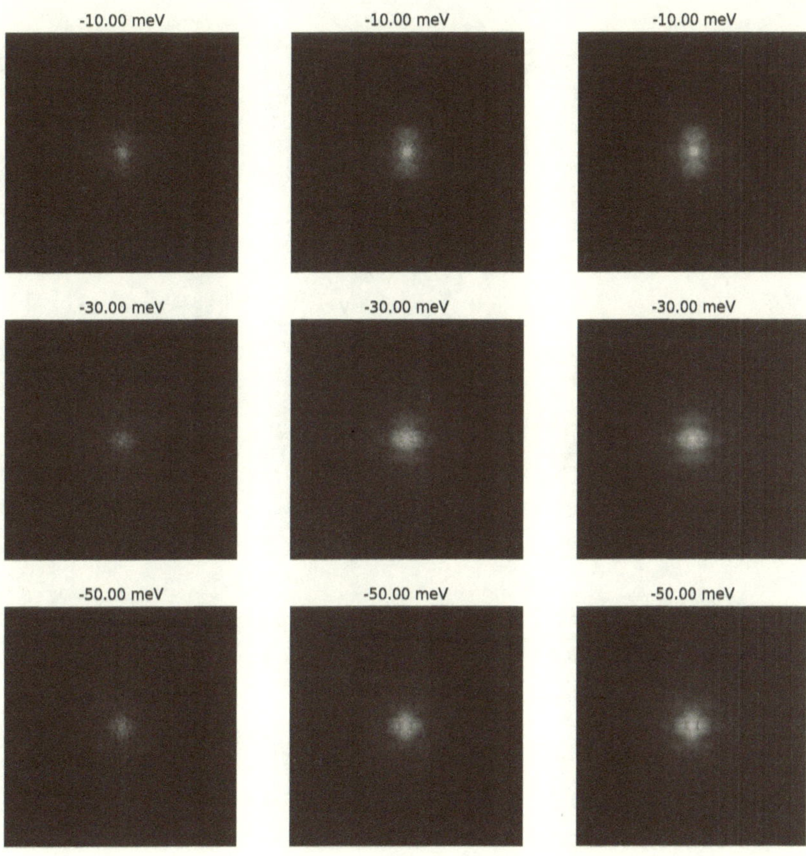

Figure A.0.2.: 2-dimensional FFT data from conductance map n. 11, measured at 4.6 K with U_{mod} =1.6 mV, U_{bias} =50 mV and I_T = 500 pA, in an area of size 60 nm $\times$ 60 nm with 256 px $\times$ 256 px. Raw data is shown at the left of the image, symmetrized at the centre and smoothed at the right for energies -10, -30 and -50 meV.

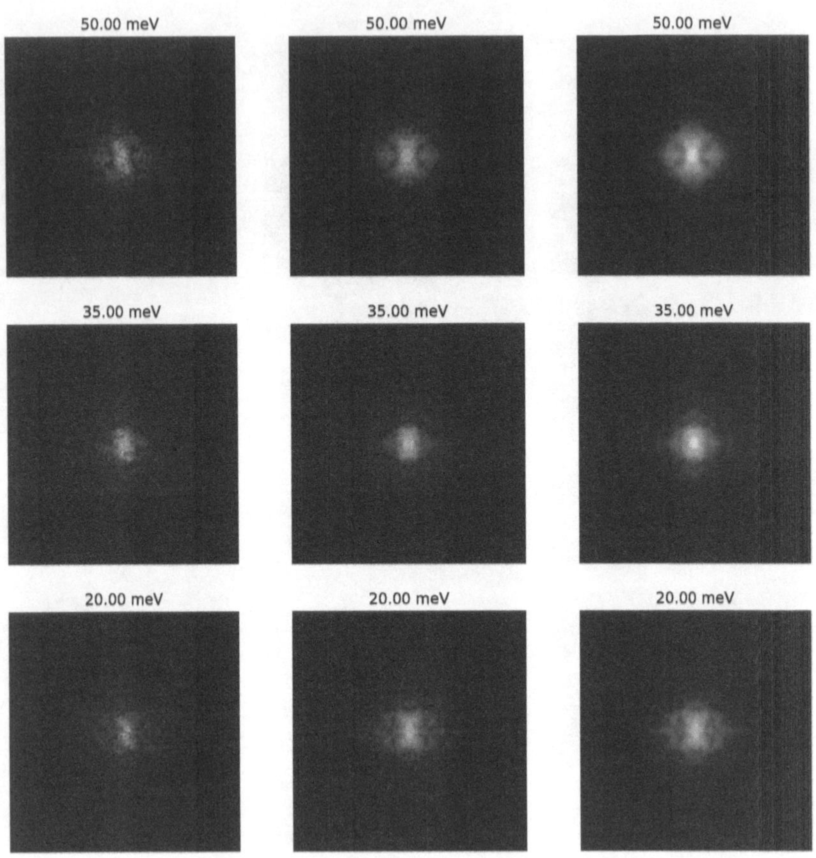

Figure A.0.3.: 2-dimensional FFT data from conductance map n. 12, measured at 4.6 K with U_{mod} =1.6 mV, U_{bias} =50 mV and $I_T = 500$ pA, in an area of size 30 nm $\times$ 30 nm with 128 px $\times$ 128 px. Raw data is shown at the left of the image, symmetrized at the centre and smoothed at the right for energies 50, 35 and 20 meV.

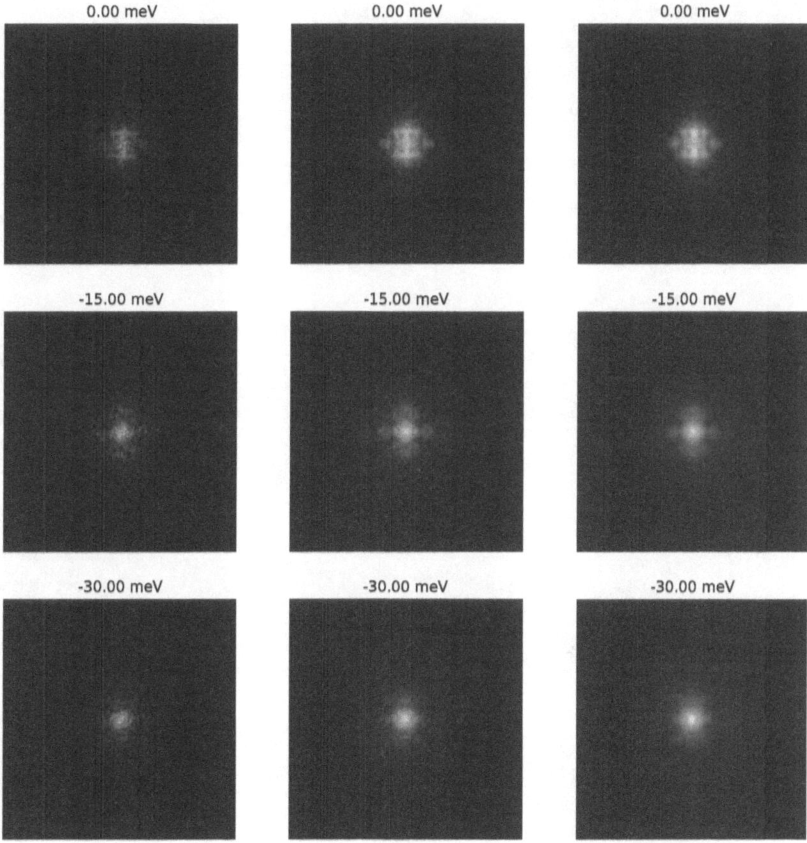

Figure A.0.4.: 2-dimensional FFT data from conductance map n. 12, measured at 4.6 K with U_{mod} =1.6 mV, U_{bias} =50 mV and I_T = 500 pA, in an area of size 30 nm $\times$ 30 nm with 128 px $\times$ 128 px. Raw data is shown at the left of the image, symmetrized at the centre and smoothed at the right for energies 0, -15 and -30 meV.

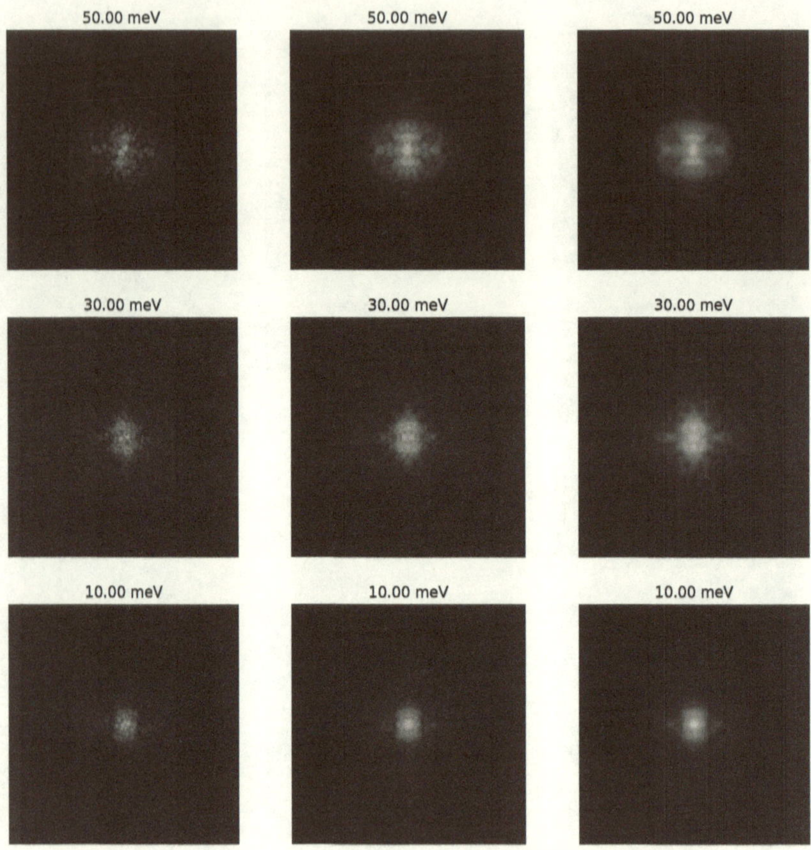

Figure A.0.5.: 2-dimensional FFT data from conductance map n. 16, measured at 12 K with U_{mod} =1.6 mV, U_{bias} =50 mV and I_T = 500 pA, in an area of size 30 nm × 30 nm with 128 px × 128 px. Raw data is shown at the left of the image, symmetrized at the centre and smoothed at the right for energies 50, 30 and 10 meV.

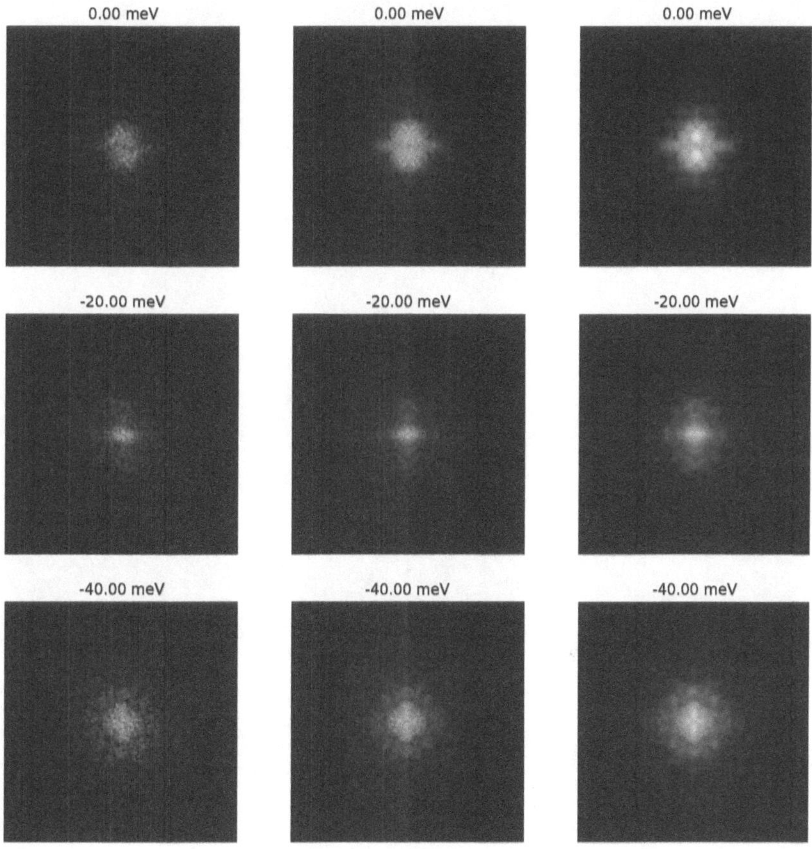

Figure A.0.6.: 2-dimensional FFT data from conductance map n. 16, measured at 12 K with U_{mod} =1.6 mV, U_{bias} =50 mV and $I_T = 500$ pA, in an area of size 30 nm × 30 nm with 128 px × 128 px. Raw data is shown at the left of the image, symmetrized at the centre and smoothed at the right for energies 0, -20 and -40 meV.

B. Na$_{95}$Li$_{0.05}$FeAs

Here are shown further studies on Na$_{1-x}$Li$_x$FeAs. In particular a temperature dependence study of Na$_{95}$Li$_{0.05}$FeAs. SI-STS maps were measured from 5.6 K up to 50 K. The data yields interesting results, however the tip states contribute to the overall tunnelling conductance changing the spectra. Therefore the observations are not conclusive.

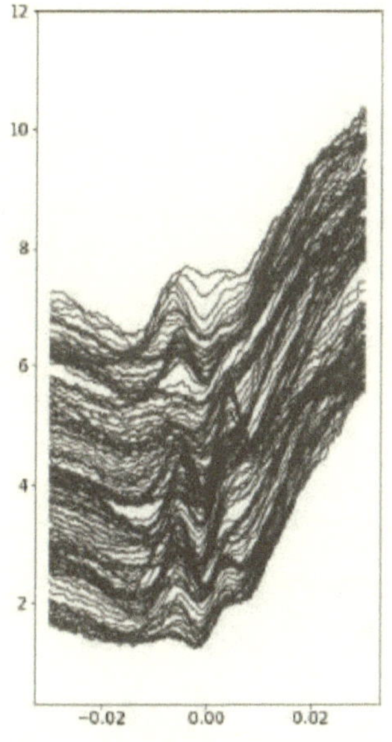

(a) dI/dU spectra taken along the red path in (b).

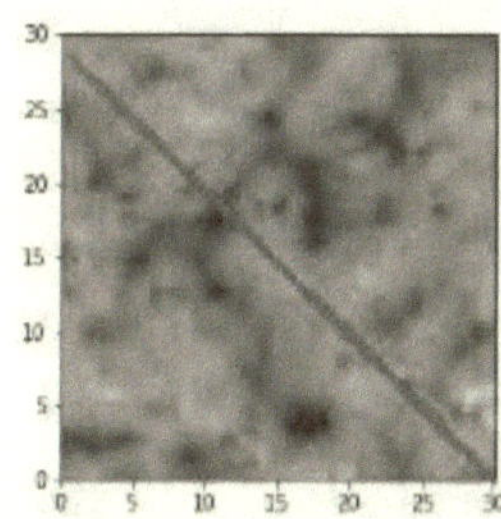

(b) 25 nm × 25 nm conductance map measured at U_{bias}= 30 mV.

Figure B.0.1.: Conductance map measured at 5.6 K with U_{mod} =1 mV, U_{bias} =30 mV and I_T = 500 pA, in an area of size 25 nm × 25 nm with 128 px × 128 px.

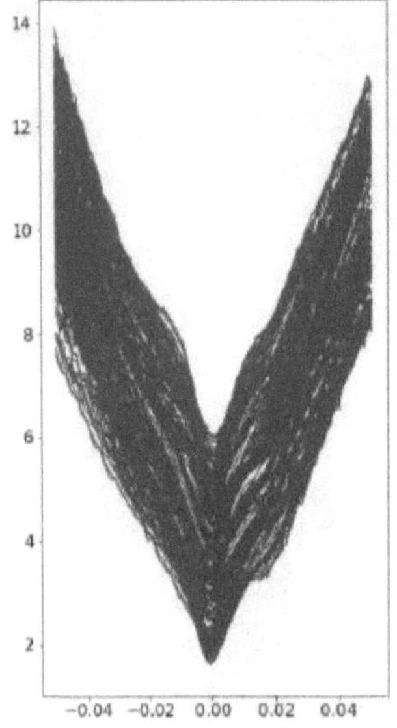

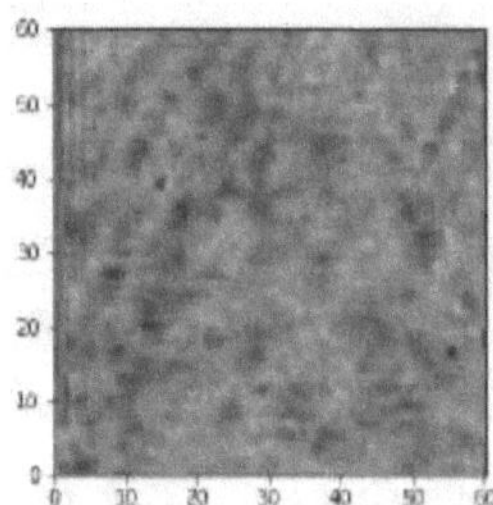

(a) dI/dU spectra taken along the red path in (b).

(b) 60 nm × 60 nm conductance map measured at U_{bias}= 50 mV.

Figure B.0.2.: Conductance map measured at 5.8 K with U_{mod} =1.6 mV, U_{bias} =50 mV and I_T = 500 pA, in an area of size 60 nm × 60 nm with 256 px × 256 px.

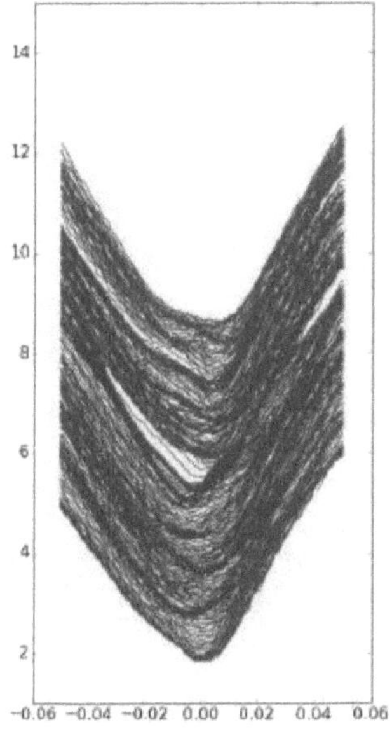

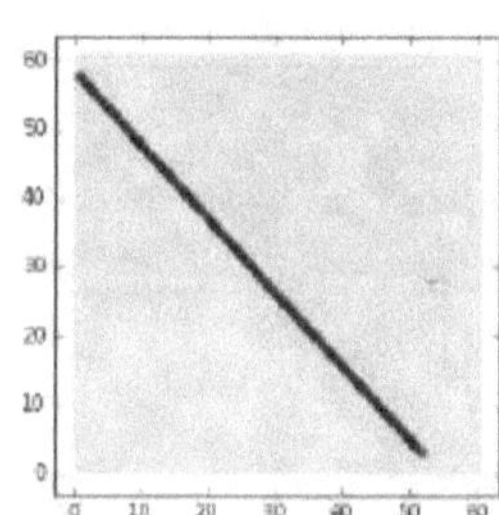

(a) dI/dU spectra taken along the red path in (b).

(b) 30 nm × 30 nm conductance map measured at U_{bias}= 50 mV.

Figure B.0.3.: Conductance map measured at 16 K with U_{mod} =1.6 mV, U_{bias} =50 mV and $I_T = 400$ pA, in an area of size 30 nm × 30 nm with 128 px × 128 px.

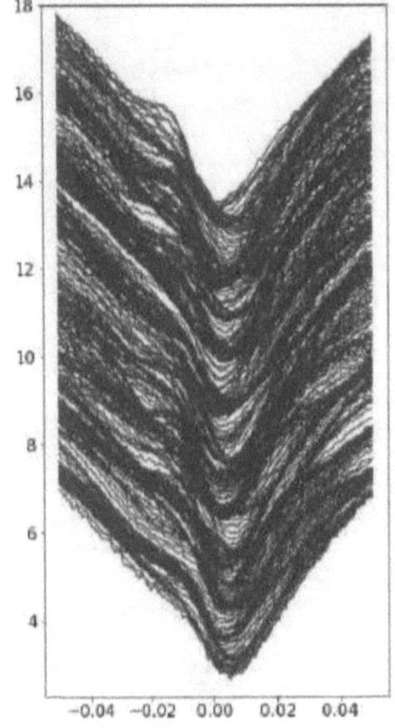

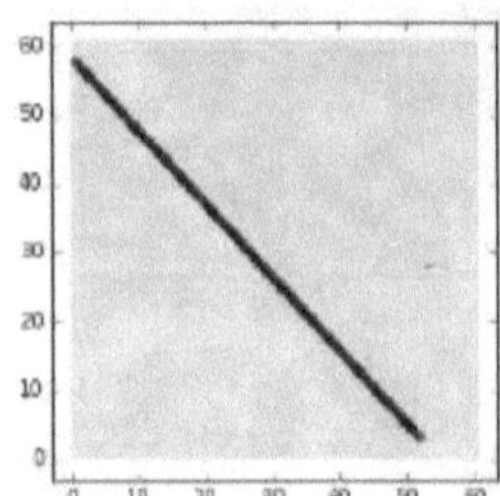

(a) dI/dU spectra taken along the red path in (b).

(b) 60 nm × 60 nm conductance map measured at U_{bias}= 50 mV.

Figure B.0.4.: Conductance map measured at 25 K with U_{mod} =1.6 mV, U_{bias} =50 mV and $I_T = 500$ pA, in an area of size 30 nm × 60 nm with 256 px × 256 px.

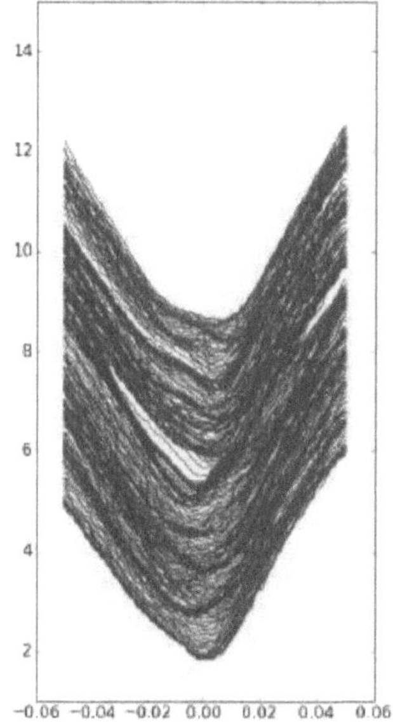

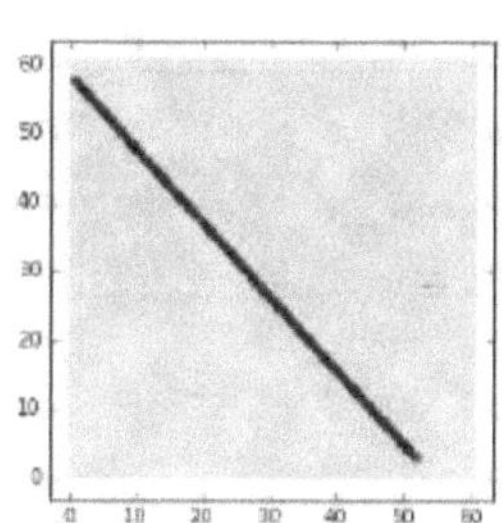

(a) dI/dU spectra taken along the red path in (b).

(b) 30 nm × 30 nm conductance map measured at U_{bias}= 50 mV.

Figure B.0.5.: Conductance map measured at 35 K with U_{mod} =1.6 mV, U_{bias} =50 mV and I_T = 400 pA, in an area of size 30 nm × 30 nm with 128 px × 128 px.

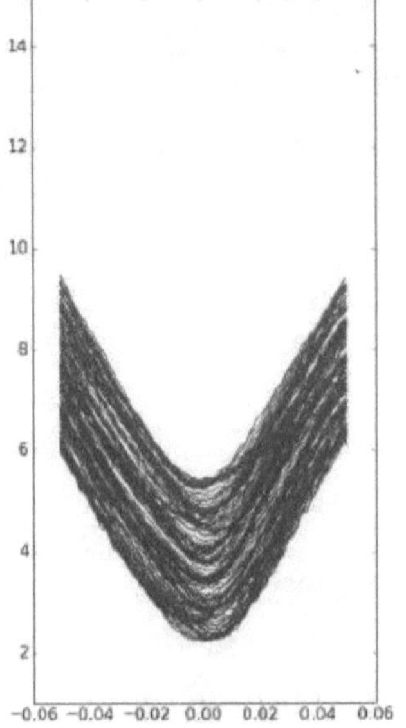

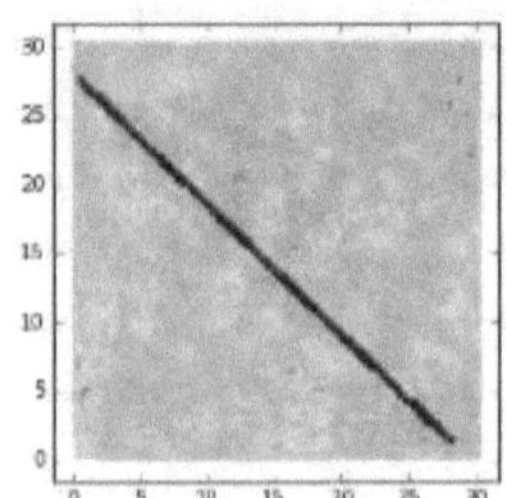

(a) dI/dU spectra taken along the red path in (b).

(b) 30 nm $\times$ 30 nm conductance map measured at $U_{bias}=$ 50 mV.

Figure B.0.6.: Conductance map measured at 35 K with U_{mod} =1.6 mV, U_{bias}=50 mV and I_T = 400 pA, in an area of size 30 nm $\times$ 30 nm with 128 px $\times$ 128 px.

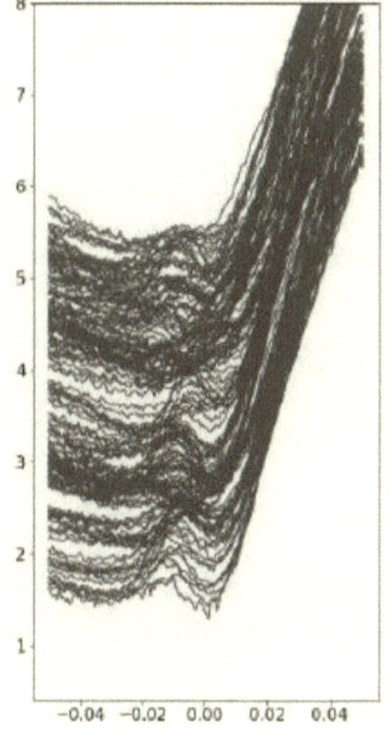

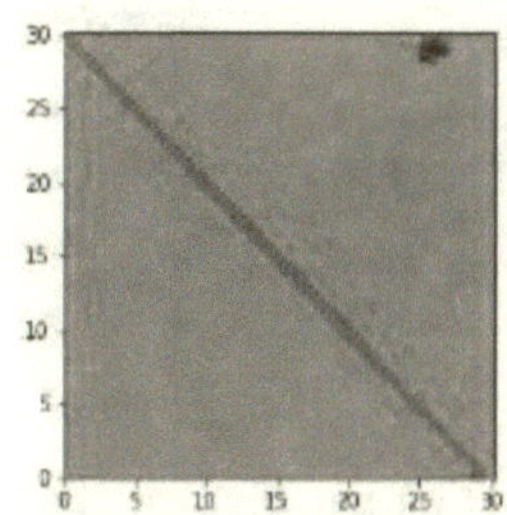

(a) dI/dU spectra taken along the red path in (b).

(b) 30 nm × 30 nm conductance map measured at U_{bias}= 50 mV.

Figure B.0.7.: Conductance map measured at 35 K with U_{mod} =1.6 mV, U_{bias} =50 mV and I_T = 400 pA, in an area of size 30 nm × 30 nm with 128 px × 128 px.

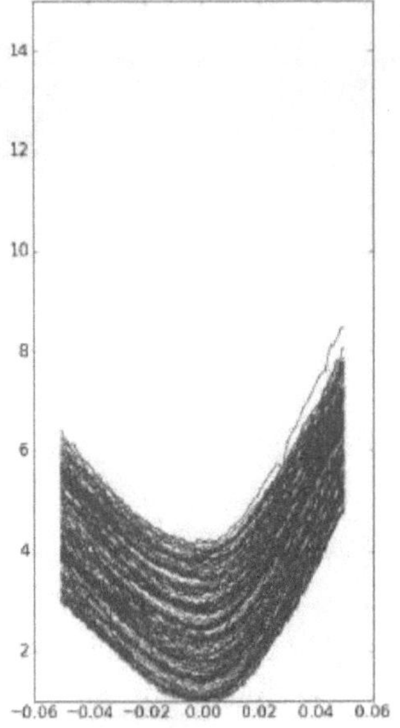

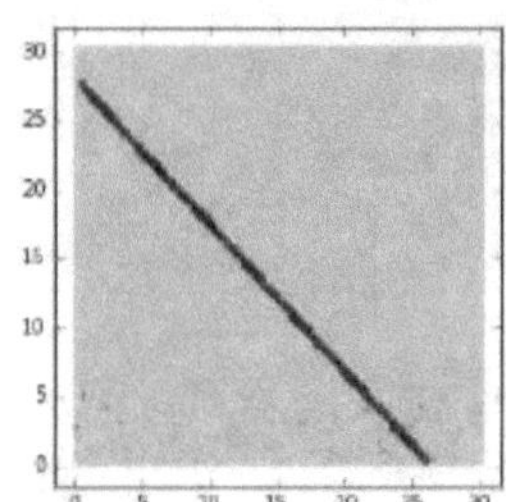

(a) dI/dU spectra taken along the red path in (b).

(b) 30 nm × 30 nm conductance map measured at U_{bias}= 50 mV.

Figure B.0.8.: Conductance map measured at 50 K with U_{mod} =2 mV, U_{bias} =50 mV and $I_T = 200$ pA, in an area of size 30 nm × 30 nm with 128 px × 128 px.

www.ingramcontent.com/pod-product-compliance
Lightning Source LLC
LaVergne TN
LVHW042158190726
843493LV00006B/1720